FEED INTO MILK

A *new applied feeding system for dairy cows*

AN ADVISORY MANUAL

Edited by

C. THOMAS

NOTTINGHAM
University Press

FEED INTO MILK

First published by Nottingham University Press

This reissued original edition published 2023 by 5m Books Ltd www.5mbooks.com

British Library Cataloguing in Publication Data
Feed into Milk: An Advisory Manual
I Thomas, C.

ISBN   9781789182811

Disclaimer

Every reasonable effort has been made to ensure that the material in this book is true, correct, complete
and appropriate at the time of writing. Nevertheless, the publishers and authors do not accept
responsibility for any omission or error, or for any injury, damage, loss or financial consequences
arising from the use of the book.

Typeset by Nottingham University Press, Nottingham

EU GPSR Authorised Representative
LOGOS EUROPE, 9 rue Nicolas Poussin, 17000, LA ROCHELLE, France
E-mail: Contact@logoseurope.eu

CONTENTS

5.   SUMMARY OF FEED CHARACTERISATION METHODS
     Intake and metabolisable energy
     Metabolisable protein
     Rumen stability value

On the enclosed CD-rom, you will find a complete copy of this book in PDF format, as well as the following:

FiM model equations in software-compatible form with glossary of terms

A guide to using the **FiM** model

Using the rumen stability DSS

Using the amino acid DSS

Using the milk quality DSS

Feed Data Base

Introduction

To be sustainable, future technological developments in livestock systems need to result in enhanced profitability together with improved predictability of output and product composition in order better to meet market needs. In achieving these aims it is vital that the welfare of the animal is not compromised and the environment is not harmed.

Feeds represent the major component of the costs of production in ruminant systems. If the efficiency of feed utilisation could be increased by 5%, and the uptake of technology were 50%, the direct benefit to the UK dairy industry would be in excess of £20 million per annum. Manipulating the diet not only influences yield and composition of milk but also can have profound effects both on the welfare of the animal and the losses of nutrients from the system which result in diffuse pollution of the environment. The Government support for strategic research in ruminant nutrition has greatly improved our understanding of the processes involved in the conversion of feed to animal product. However, the lack of a framework to incorporate this knowledge into support systems which enable both the dairy farmer and the policy maker to predict the consequences of change has meant that much of the technology has not been transferred into practice in a way that can aid the sustainability of milk production systems.

The aim of Feed into Milk (FiM) was to develop this framework and to derive an improved nutritional model that could be applied in advisory practice with almost immediate effect. The project comprised reviews and evaluations of current relevant information on feeding systems, a joint modelling, animal experiment and feed evaluation approach to deriving an improved diet formulation system and the construction of an improved applied nutrition model to meet the future needs of the dairy farmer and the feed industry.

The project was funded through the *LINK* Sustainable Livestock Production programme by Defra, DARDNI and SEERAD (50%), The Milk Development Council and AgriSearch (25%) and 32 commercial companies from the dairy support industry (25%). A consortium of the funders managed the project. Consortium meetings were held every six months to agree the future programme using a participative approach. ADAS, ARINI and SAC were the main contractors (with subcontract to CEDAR).

Dissemination of the information from the project was a key priority and this was assisted by the consortium through press releases and farmer friendly articles, a Feed into Milk website, the preparation and circulation by the Milk Development Council of 35,000 copies of a Farmers Booklet together with a simple illustrative computer dairy feeding program of the system.

This advisory manual represents the next stage in this process. It is aimed at consultants, research workers, academics and students of animal nutrition and is in two parts in separate formats (text and CD). The first (text) part explains the background and principles involved in deriving the FiM system and provides the rationale and the equations for the prediction of intake and the calculation of the requirements and supply of energy and protein. In addition the manual recognizes that an applied feeding system is only part of the process of diet formulation and provides a series of decision support systems (DSS) to assist in building rations for dairy cows.

The accompanying CD explains the principles of diet formulation and provides examples on the use of the DSS. The complete set of equations is listed in software format and this together with an extensive feed database enables the user to begin formulating diets with almost immediate effect.

LIST OF FUNDERS AND CONSORTIUM MEMBERS

Government Sponsors

Defra, DARDNI and SEERAD through the *LINK* Sustainable Livestock Production Programme

Levy Boards

Milk Development Council
AgriSearch

Dairy support industry

ADAS Consulting Ltd

Agricultural Research Institute of Northern IrelandI

ABNA (formerly J Bibby Agriculture Ltd.)

Aventis (formerly Rhone Poulenc Animal Nutrition)

BOCM PAULS

Carrs Billington Agriculture Ltd (includes AF plc, Billington Agriculture Ltd, Billington Feeds Ltd, Carrs Agricultural Ltd)

Countrywide (includes MSF Ltd, RM Formulations, WMF Ltd, SCATS Ltd)

Dalgety Agriculture Ltd

Diageo Global Supply (formerly United Distillers)

Dynamic Nutrition Services Ltd

Ecosyl Products Ltd

Farmway Ltd

Frank Wright Ltd

FSL Bells

Heygate & Sons Ltd

John Thompson & Sons Ltd

Massey Bros (Feeds) Ltd

Mole Valley Farmers

North Eastern Farmers

Nutrition Services (INT) Ltd

NWF Agriculture

Promar International (formerly Genus Management)

Provimi Ltd (formerly Nutec Ltd)

Pye Farm Feeds (W&J Pye Ltd)

Rumenco

SAC (including DRC and Signet)

Scottish Livestock Services

UK Association of Fish Meal Manufacturers

UM Group

Wynnstay & Clwyd Farmers plc

FEED INTO MILK – PROJECT MANAGEMENT AND DELIVERY

Consortium Chairman	Dr. John Allen, Frank Wright Ltd
Technical Secretary	Dr. Jonathan Blake, Dynamic Nutrition Services Ltd
Project Coordinator	Prof. Cledwyn Thomas, SAC

Contractors

ADAS	Dr. Bruce Cottrill
	Prof. Ian Givens
	Dr. Caroline Rymer
Agricultural Research Institute of Northern Ireland (ARINI)	Dr. Rosemary Agnew
	Dr. Tim Keady
	Dr. Sinclair Mayne
	Dr. Tianhai Yan
Scottish Agricultural College (SAC)	Dr. Nick Offer
	Dr. Dawn Percival

Subcontractors

CEDAR, University of Reading	Prof. Geoff Alderman
	Prof. David Beever
	Prof. Jim France
	Dr. Ermias Kebreab

ACKNOWLEDGEMENTS

I wish to record my thanks to Dr. J. Newbold for his invaluable comments on the manuscript, to Ms L. Gechie for administrative and secretarial support and to Prof. F. J. Gordon for overall guidance. S. Gilkinson, R. Mansbridge, R. Lawson and M. Patterson made valuable contributions to the prediction of voluntary intake and the estimation of energy requirements was greatly supported by Dr. E. F. Unsworth, S. Cammell, and M. Porter. For assistance in the techniques for the characterization of feeds I wish to acknowledge Dr. R. S. Park and all the staff of ADAS Nutritional Sciences Research Unit.

I am grateful for the support of the contractors and to the leadership and enthusiasm of Dr. John Allen as Chairman of the Consortium. Dr. Jonathan Blake's contribution as Technical Secretary of the project was invaluable and I am particularly grateful for his continued role in the production of this manual. Finally, I would like to thank all the members of the consortium, not only for their financial support but also for their contribution to the scientific direction of the project.

I would like to dedicate this manual to the memory of Geoff Alderman to acknowledge his contribution to the development of feeding systems in the UK and to their incorporation into practice.

1   Prediction of Voluntary Intake

T.W.J. Keady, S. Mayne, N.W. Offer and C. Thomas

Background

Accurate prediction of feed intake is a fundamental prerequisite of any nutritional model designed to provide feeding recommendations for lactating dairy cattle. Consequently, much research effort has been expended over the last 30 years in developing prediction models. These range from relatively simple multiple regression equations to more complex models embracing animal, food and environmental factors. However, given the major changes in the types of diet now offered to dairy cows, coupled with progress in genetic merit/milk production potential, it is important to examine if the feed intake prediction models currently available for use are appropriate for today's dairy cow. A key objective of **FiM** was to examine the performance of existing prediction models for feed intake and, if necessary, to develop new models that could cope better with modern cows and production systems.

Equation VH1 from Vadiveloo and Holmes (1979) has been widely recommended for advisory use (TCORN report No 5, AFRC, 1990 and AFRC, 1993)

$$TDMI = 0.076 + 0.404\ CDMI + 0.013\ W - 0.129\ WL + 4.12\ \log WOL + 0.14\ Y$$

where TDMI is the total dry matter intake (kg/day); CDMI, concentrate dry matter intake (kg/day); W, live weight (kg); WOL, the week of lactation and Y, milk yield (kg/day).

Recognising that equation VH1 takes no account of the effects of forage quality on intake, AFRC (1993) also quote the equation of Lewis (1981). This included variables describing the digestibility and fermentation quality of grass silage. In France, INRA (Dulphy *et al.*, 1989) developed a system based largely on forage characteristics (defined as fill units) whilst, in contrast,

Milligan *et al.*, (1981) derived an equation based solely on animal and environmental variables. More recently a MAFF funded project 'RUMINT' (Oldham *et al.*, 1998) recommended an approach to consider both feed and animal factors by distinguishing between diets that allow intakes to be unconstrained (essentially a maximum intake to meet cow energy needs) with those that are constrained by their physical or chemical characteristics.

The ability of these published equations to predict intake was evaluated against a database constructed from 27 dairy cow studies undertaken at ARINI, ADAS and SAC in which 2425 individual cows were offered grass silage as the sole forage. This was part of a larger database of 3337 cows in which other conserved forages were included (see later and Appendix 1.1 for details).

Table 1.1. Prediction of the total dry matter intake (TDMI, kg/d) of grass-silage based diets by a range of equations

Equation	TDMI Observed = 16.96		MSPE	Proportion of MSPE		
	Predicted	Bias		Bias	Line	Random
VH1	16.99	-0.03	2.9	0.00	0.00	1.00
Lewis (1981)	16.08	0.88	5.5	0.14	0.00	0.86
Milligan *et al.* (1981)	19.14	-2.18	8.9	0.54	0.02	0.44
Oldham *et al.*(1998)	19.02	-2.06	10.8	0.39	0.17	0.44
Dulphy *et al.* (1989)	19.84	-2.88	11.4	0.73	0.01	0.26

The evaluation was confined to grass silage diets since the equations that included feed characteristics were largely derived from this base. Nevertheless, the results showed a marked variation in the accuracy and precision of prediction. The VH1 equation of Vadiveloo and Holmes, (1979) had the lowest bias and mean square prediction error (MSPE). The equations of Oldham *et al.* (1998), Milligan *et al.* (1981) and Dulphy *et al.* (1989) over-predicted intake by 12, 13 and 17% respectively, whilst Lewis (1981) under-predicted intake by 5%.

Principles: Towards an improved prediction equation

Although equation VH1 performed well in the evaluation outlined above, it does not include as variables those factors that are known to influence the voluntary intake of silage-based diets (Osbourn and Thomas, 1989; Steen *et al.*, 1998). These include the digestibility and the fermentation quality of the silage and the concentration of protein in the concentrate. In particular the

VH1 equation does not take into account that the response in total intake to concentrate is closely related to the intake of the forage as the sole feed (forage intake potential) and further that the response is curvilinear in nature (Osbourn and Thomas, 1989).

Clearly any new equation must be at least as accurate as VH1 but also it must be biologically meaningful and be relevant to a wider range of forages than grass silage. It will need to describe:

- the intake potential of a forage as the sole feed (Forage Intake Potential, FIP)

- the interaction between FIP and the amount of concentrate in the diet.

- the curvilinear relationship between the intake of forage (and total diet) and the amount of concentrate in the diet

- the composition of the concentrate

- the effect of cow variables such as:

  - stage of lactation

  - live weight, condition score

  - milk yield.

FiM intake equation

Methodology

Prediction of forage intake potential (FIP)

Conserved forages are rarely offered to dairy cows as the sole feed and thus a methodology was needed to estimate FIP (intake of forage as the sole feed) from beef cattle studies and from experiments where dairy cattle had been given mixtures of forage and concentrate.

A reference set of data was compiled from 136 grass silages given as the sole feed to beef cattle at ARINI and 34 grass silages from SAC that were offered to dairy cows supplemented with 7 kg/d of concentrate and yielding, on average, 28 kg of milk per day. These data were transformed into a reference set of 170 intake values standardised to zero concentrate intake and an equalised milk yield (i.e. FIP).

This transformation involved two steps:

1.   Conversion of ARINI data for intake of beef cattle (at zero concentrate intake) to intakes expected from dairy cattle (at zero concentrate intake), and

2.   Conversion of SAC data for intake of dairy cattle (at 7kg/d concentrate intake and 28kg milk/d) to intakes expected from dairy cattle at zero concentrate intake and at a standard yield of milk.

In Step 1, an independent dataset from ARINI comprising 13 grass silages offered as the sole feed to both beef cattle and dairy cows (average yield 8 kg/d) was used to generate a regression equation (see Appendix 1.2) that was then applied to the beef cattle subset of 136 grass silages to derive the FIP values for dairy cows.

In Step 2, an adjustment for concentrate intake was derived from an independent study undertaken at SAC in which 8 grass silages were each offered at 4 levels of concentrate ranging from 0 to 12kg (Appendix 1.2). Milk yield at zero concentrate in this study was 8kg/d, a value similar to that observed in Step 1 and this was taken to be the standard milk yield at zero concentrate for the definition of FIP. A milk yield adjustment of 0.1486Y was then derived from a review of the literature and this together with the concentrate adjustment was applied to the 34 SAC values in the reference dataset to produce the FIP values (at zero concentrate and a daily yield of 8kg milk).

The combined reference set of 170 values (FIP, g/kg $W^{0.75}$) was then used to develop a Near Infra Red (NIR) prediction model based on scans of the grass silage samples relating to the reference values (see later section on Feed Characterisation Methods). The overall model was tested against two sets of independent data and resulted in R^2 of 0.69 and 0.76 (Agnew *et al.*, 2001). FIP values for forages other than grass silages were calculated from their chemical composition and digestibility. The equations are presented in the section on Feed Characterisation Methods.

Derivation of the FiM intake equation

To derive the new equation, a database was compiled from 3337 observations of lactating dairy cows comprising a wide range of diets and feeding systems relevant to modern dairy production systems. This was constructed from data involving cows given grass silage (n = 3136), maize silage (n = 161) or whole crop wheat based diets (n = 40) (see Appendix 1.1 for detail). Total DM intake varied from 8.5 to 29.4 kg/d (Concentrate 1.5 to 21.4 kg DM/ d) and daily milk yields ranged from 7.7 to 49.7 kg. A variety of multi-variate methods including stepwise multiple regression, best subsets multiple regression, partial least squares regression and principal components regression were used to derive prediction equations from the dataset. A subset based on information from SAC Langhill, where dairy cows were given diets of fixed composition for up to 46 weeks of lactation, was used to develop a stage of lactation adjustment factor.

Calculation of intake

The multi-variate methods outlined above accounted for very similar proportions of the variance in the observed intake of total DM (mean = 17.2 kg DM/d). However an equation based on a stepwise multiple regression (SMR) analysis is recommended for two reasons. Firstly SMR yielded the most biologically sensible coefficients and secondly, it performed better than others when tested against an independent database (see later). Total DM intake (TDMI, kg/d) is calculated from

$$TDMI = -7.98 + 0.1033 \text{ FIP} - 0.00814 (\text{FIP*CDMI}) - 1.1185 \text{ CS} + 0.01896 W + 0.7343 \text{ CDMI} - 0.00421 (\text{CDMI})^2 + 0.04767 \text{ E}_1 - 6.43 (0.6916^{WOL}) + 0.007182 [\text{FS}] + 0.001988 ([\text{CCP}]*\text{CDMI})$$ (Equation 1.1)

where FIP is the forage intake potential (g/kg $W^{0.75}$); CDMI, the concentrate dry matter intake (kg/d); CS the condition score of the cow(1-5 scale); W the live weight (kg) E_p milk energy output (MJ/cow/d); WOL, week of lactation (constrained to maximum of 10) [FS], forage starch concentration (g/kg DM) and [CCP], the crude protein concentration of the concentrate (g/kg total concentrate DM).

Where a forage mixture is offered, FIP and FS are weighted values calculated according to the relative proportions of each forage on a DM basis.

Evaluation of the FiM intake equation

The accuracy and precision of the **FiM** intake equation was evaluated using two independent datasets that were compiled from data in which cows had been offered either grass silage as the sole forage or mixed forage-based diets (see Appendix 1.3 for details). The grass silage-based diets (n = 34) represented a wide range of food and animal variables with total intake varying from 9.8 to 20.2 kg DM/d. However, the range in variables for the mixed forage-based diets was not as large (16.4 to 19.8kg DM/d), due to the small number of studies available (n = 10). Plots of actual intake against those predicted by the **FiM** equation are shown in Figure 1.1.

a) Grass silage diets (n=34)

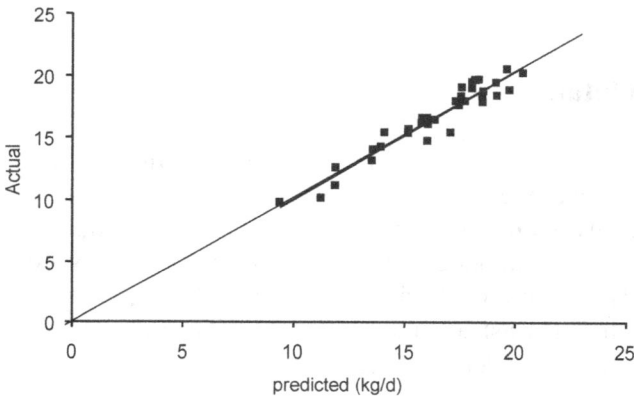

b) Mixed forage diets (n=10)

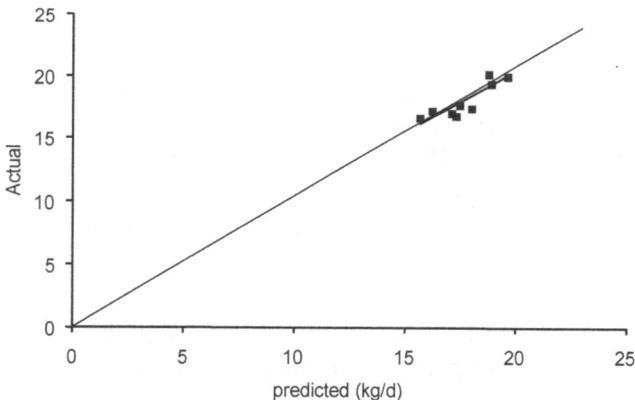

Figure 1.1 Evaluation of the **FiM** intake equation- (Total DM Intake, kg/d)

The validation statistics in Table 1.2 clearly illustrate that, regardless of the basal forage, the **FiM** intake equation was accurate and precise at predicting feed intake. It underpredicted the intake of grass silage by only 1.8% and overpredicted the intake of mixed forages by 1%. In contrast, equation VH1 of Vadiveloo and Holmes, (1979), representing the best of the equations that were evaluated earlier (see Table 1.1), over-predicted the intake of grass silage-based diets by 4% and under-predicted the intake of mixed forage-based diets by 8%. The prediction presented in the list of equations on the CD has been adjusted for a bias of 0.3 kg DM/d.

Table 1.2 Comparison of the accuracy and precision of the FiM and VH1 (1979) intake equations for grass silage (n=34) and mixed forage based based diets (n=10)

Model	TDMI (kg/day)				Proportion of MSPE		
	Actual	Predicted	R^2	MSPE	Bias	Line	Random
Grass silage							
FiM	16.6	16.3	0.93	0.658	0.08	0.05	0.87
VH1	16.6	17.2	0.87	1.480	0.24	0.05	0.71
Mixed forage							
FiM	18.0	18.2	0.75	0.519	0.09	0.01	0.91
VH1	18.0	16.5	0.62	3.020	0.70	0.06	0.24

The recommended equation contains a concentrate CP intake (CCP*CDMI) term. This term was significant in the SMR model and highlights the well-recognised positive effect of concentrate CP on silage intake. However, extrapolating beyond the limit of CP intake from which the model was constructed exaggerates the effect of CP on intake. Thus, as safety measure, it is recommended that the model be constrained to a maximum value of 3500 g concentrate CP/d.

The inclusion of the model into some linear programming systems was shown to cause problems of excessive iteration. An alternative to the recommended equation (1) was derived excluding the term ([CCP]*CDMI) and the following can be used when this problem occurs.

$$TDMI = -7.38 + 0.1018 \, FIP - 0.00795 \, (FIP*CDMI) - 1.065 \, CS + 0.01929 \, W + 0.954 \, CDMI + 0.00364 \, (CDMI)^2 + 0.05204 \, MEO - 6.894*(0.6932^{WOL}) + 0.010747 \, FS$$

(Equation 1.2)

Evaluation of equations that omitted a CP term showed little effect on bias but an increase in MSPE of 0.09.

Conclusions

The **FiM** Intake equation is an advance on the previous equations recommended for advisory use.

*   Provides a biologically sound basis for the prediction of the intake of conserved forage diets.

*   Predicts intake of grass silage-based diets more accurately and precisely than the previous widely recommended equations.

*   Facilitates accurate prediction of the intake of mixed forage diets.

*   Contains a lactation adjustment that accurately predicts the increase in intake over early lactation.

*   Applicable to a wide range of modern feeding systems (e.g. TMR).

Appendix 1.1 Range in chemical composition of silage and concentrates, food intake and animal performance (n = 3337) in the **FiM** intake dataset

	Minimum	*Maximum*	*Mean*	*s.d.*
Grass silage composition				
Alcohol corrected toluene DM (g/kg)	170	479	271	68.2
pH	3.49	5.27	4.00	0.32
Crude protein (g/kg DM)	105	213	157	22.6
Ammonia N (g/kg N)	37	200	90	32.6
Acid detergent fibre (g/kg DM)	265	420	326	31.8
DOMD (g/kg DM)	551	787	715	43.4
Maize silage composition				
Alcohol corrected toluene DM (g/kg)	226	390	323	45.0
Crude protein (g/kg DM)	74	101	89	9.3
Starch (g/kg DM)	114	366	288	62.9
Whole crop wheat				
Alcohol corrected toluene DM (g/kg)	301	584	473	106.0
Crude protein (g/kg DM)	104	199	132	39.3
Starch (g/kg DM)	32	306	170	98.1
Concentrate composition (g/kg DM)				
Crude protein	135	360	232	39.9
Starch	9	505	250	89.6
Acid detergent fibre (g/kg DM)	54	170	104	23.7
Food intake (kg DM/cow/day)				
Grass silage	1.6	18.8	8.9	2.62
Maize silage	4.8	12.3	8.6	1.8
Whole crop wheat	5.8	11.3	8.0	1.0
Straw	0.3	2.7	0.9	0.5
Concentrate	1.5	21.4	7.5	2.96
Brewers grains	0.4	2.2	0.9	0.30
Total	8.5	29.4	17.2	3.1
Milk yield (kg/day)	7.7	49.7	26.5	6.96
Milk composition				
Fat (g/kg)	15.3	65.3	41.4	6.01
Protein (g/kg)	24.3	46.2	32.0	2.81
Live weight (kg)	351	851	588	69.8
Week of lactation	2.0	44	17.1	6.8

Appendix 1.2 Equations used to generate FIP values

$$y = 20.90 + 0.906\ (0.1616)^x \quad R^2 = 0.74 \quad P < 0.001$$

where y is the intake of silage as a sole feed by dairy cows (g DM/kg$^{0.75}$) and x the intake as a sole feed by beef cattle (g DM/kg$^{0.75}$).

$$SDMI_c = 1.191\ SDMI_0 - (0.191\ SDMI \times 1.013^c)$$

where SDMI$_c$ is the silage intake at concentrate intake c (g DM/kg$^{0.75}$); SDMI$_0$ is the silage intake at zero concentrate intake (g DM/kg$^{0.75}$) and C the concentrate intake (g DM/kg$^{0.75}$).

Appendix 1.3 Range in food and animal parameters in the independent dataset for grass silage- (n = 34) and mixed forage- (n = 10) based diets

	Minimum	Maximum	Mean
Grass silage diets			
Forage intake potential (g/kg W$^{0.75}$)	80	109	93
Concentrate intake (kg DM/day)	0.8	12.9	7.9
Condition score	2.0	3.4	2.7
Live weight (kg)	505	628	563
Milk energy output (MJ/day)	46	112	86
Actual intake (kg DM/day)	9.8	20.2	16.5
Mixed forage diets			
Forage intake potential (g/kg W$^{0.75}$)	84	110	95
Concentrate intake (kg DM/day)	5.2	6.0	5.9
Condition score	2.5	2.9	2.6
Live weight (kg)	565	597	582
Milk energy output (MJ/day)	67	95	84
Actual intake (kg DM/day)	16.4	19.8	18.0

2   Energy Requirement and Supply

R.E. Agnew, T. Yan, J. France, E. Kebreab and C. Thomas

Background

The UK metabolisable energy (ME) feeding system, developed by Blaxter (1962), was first proposed for use in the UK by the Agricultural Research Council (ARC, 1965). Later, a simplified system was recommended for advisory use and published as Technical Bulletin 33 (MAFF, 1975). The original ME system was later revised as ARC (1980) and further modified by the Agricultural and Food Research Council (AFRC, 1990) with a new working version for advisory use being published in AFRC (1993). In this (as in the previous versions) the ME value of a feed is defined as being measured at the maintenance level of feeding. The ME consumed (MJ/d) is calculated as:

$$ME = GE - FE - UE - MethE$$

where GE is the gross energy (heat of combustion) of the feed consumed, and FE, UE and MethE is the energy lost in faeces, urine and methane respectively.

The net energy (NE) is that part of the feed ME consumed that is used by the animal for maintenance and production:

$$NE = ME \times k$$

where k is the efficiency of utilization of ME for the relevant metabolic process.

Over the same period, a number of net energy (NE) systems were being developed in Europe (e.g. Van Es, 1978) and North America (NRC, 1988). Although the systems share many of the principles, the UK ME model is unique in that feed energy values can be stated independently of the process for which they are used making it easy to tabulate values and compare feeds. Further the system is factorial in construction, whereas the NE systems are regression models, where maintenance is the calculated intercept of a regression of energy balance on energy inputs measured at the production level.

11

Principles: The FiM energy system

The ME system remains a sound basis for rationing dairy cows and further there is an extensive methodology to enable feeds to be characterised in terms of their ME content. However, as pointed out by Agnew and Yan (2000), the system needs to be updated and modified to take account of

- Requirements that are more representative of those of modern high genetic merit dairy cows.

- Diets that are more representative of those normally consumed by dairy cows within the UK.

- Recent reports of the ME requirement for maintenance showing higher values than those previously recommended.

- The widespread availability of computer based rationing systems that obviate the need for a factorial system as a simplification of the curvilinear relationship between energy intake and output.

Over the last 10 years a substantial number of calorimetric measurements using high yielding dairy cows given diets representative of UK feeding practices have become available. These have been undertaken in the main at ARINI and CEDAR. However, separate analysis of the data from the two centres (Yan *et al.*, 1997a and Cammell *et al.*, 1998) resulted in very different estimates of the ME required for maintenance (M_m) and for the efficiency of utilisation of ME for lactation (k_l) (ARINI, M_m = 0.67 MJ/kgW$^{0.75}$ and k_l = 0.65; CEDAR, M_m = 0.51 MJ/kg logW$^{0.75}$ and k_l = 0.55). No clear reasons emerged as to the cause of the differences since methodologies were essentially the same. A collaborative project between ARINI and CEDAR was initiated with the aim of resolving the differences and deriving a new approach through

- combining the calorimetric information available and resolving any differences in the interpretation of data

- determining new relationships between output and supply that more adequately reflect the biology of energy transactions in the dairy cow.

Modelling the relationship between intake and output to derive ME requirement

The calorimetric data from ARINI and CEDAR were combined with others from the Grassland Research Institute, Hurley and DARD/Queen's University, Belfast to provide a database from which to derive new relationships. A total of 642 individual cow records (Appendix 2.1) were collated and verified from all four centres.

A new empirical modelling approach was developed to interpret the calorimetric data (Kebreab *et al.*, 2003). This differs from the factorial approach previously used by ARC (1980) and AFRC (1993). Instead of deriving maintenance requirements from measurements of fasting metabolism, the new method allowed simultaneous calculation of ME_m and k_l. It is based on modelling the relationship between milk energy output (E_l, measured from the product of milk yield and the gross energy concentration of the milk) and measured ME input (MEI with full measurement of losses in faeces, urine, methane, and heat).

The following definition of k_l was adopted:

$$k_l = \frac{\text{milk energy derived from diet MEI}}{\text{diet MEI directed towards milk production}}$$

Thus:

- For cows in positive energy balance, when some ME intake is used for growth, MEI is adjusted.

- When cows are in negative energy balance, some body energy is used to support lactation so E_l is adjusted.

To make these adjustments it is necessary to determine k_t (the efficiency with which body energy is used to support milk production in negative energy balance) and k_g (the efficiency with which feed ME is used to support body energy gain). Values for k_t and k_g were derived from the data independently of k_l by an iterative procedure. Firstly, values of MEI were plotted against E_l for all cows in zero energy balance (+/- 5 MJ/d). Values for cows in negative energy balance were then added to the plot with E_l values corrected using a range of values for k_t. The value for k_t that caused the regression line (MEI/ E_l) for cows in negative balance to be closest in slope and intercept to that for cows at zero energy balance was adopted as the correct estimate of k_t. The same procedure, but adding data for cows in positive energy balance, was used to estimate k_g.

Once values for k_t and k_g had been obtained the mathematics of the relationship between MEI and $E_{l(0)}$ (E_l corrected for energy balance) could be investigated. Five functions were evaluated including one linear and four non-linear methods that were developed by Kebreab *et al.* (2003) (Appendix 2.2).

The residual sum of squares and the variation explained by fitting the functions to the data were similar across all the five models that were examined. However the Mitscherlich relationship ($R^2 = 0.85$) for which the fasting metabolism value was constrained to a measured value of 0.453 MJ/kgW$^{0.75}$ (Yan *et., al.*, 1997b) is recommended for use since it provides the best biological description of the relationship between milk energy output derived from MEI and MEI directed towards milk production

$$E_{l(0)} = 5.06 - (5.06 + 0.453)e^{(-0.1326 * MEI)}$$

where $E_{l(0)}$ is adjusted milk energy output (MJ/kgW$^{0.75}$) and MEI is the adjusted metabolisable energy intake(MJ/kgW$^{0.75}$).

The relationship (see Figure 2.1) generates the following:

• Fixed ME requirement for maintenance = 0.647 MJ/kgW$^{0.75}$

• An efficiency of use of ME for lactation (k_l) that varies with MEI.

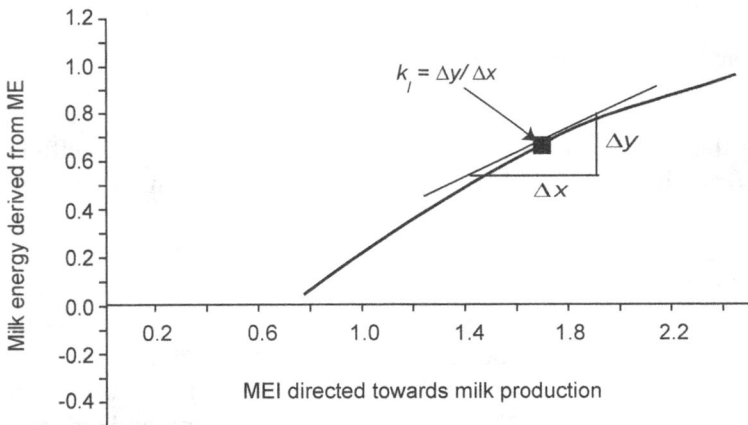

Figure 2.1 Relationship between milk energy derived from ME intake (MEI, MJ/kgW$^{0.75}$) and MEI directed towards milk production (MJ/kgW$^{0.75}$)

Neither diet effects (e.g. proportion of forage in the diet, type of forages) nor cow characteristics (e.g. stage of lactation, condition score) had any significant effect on the above relationship. Nevertheless the relationship is capable of being modified as further information becomes available on diet and cow effects.

Calculation of ME requirement

Total ME^{FiM} requirement (M, MJ/d) is defined as the product of the ME needed for:

- body weight gain, (M_g)

- pregnancy post 250 days, (M_c)

- maintenance and milk production, (M_{ml})

- activity (M_{act}).

The equations needed to derive total ME requirement are given below. Firstly the ME required for live weight gain or the net energy available for milk production from live weight loss is calculated.

Live weight gain

The ME needed for weight gain (i.e. WC (weight change, kg/d) > 0) is calculated as:

$$M_g^{FiM} = ([EV_g]*WC)/k_g \qquad \text{(Equation 2.1)}$$

where [EV$_g$] is the net energy value of weight gain (MJ/kg) and k$_g$ (the efficiency of utilisation of ME for gain) is that derived from modelling the relationship between intake and output.

A value of 19.3MJ/ kg is assumed for [EV$_g$] (AFRC, 1993), k$_g$ was derived as being 0.65 and thus:

$$M_g^{FiM} = 19.3 * WC /0.65 \qquad \text{(Equation 2.2)}$$

Live weight loss

The net energy used for milk production from weight loss (E_{lWC}, MJ/d, WC<0), is calculated from:

$$E_{lwc} = 19.3 * WC * 0.78 \qquad \text{(Equation 2.3)}$$

where again net energy value of weight loss is 19.3 MJ/kg and the efficiency of utilisation of body energy for lactation (k_t) derived from modelling the relationship between intake and output is 0.78.

Pregnancy

ME requirements for pregnancy are as described in AFRC (1993) and the detail is presented in the full list of equations on the CD.

Maintenance and milk production

The ME required for maintenance and milk production (M_{ml}, MJ/kgW$^{0.75}$) is derived from the Mitscherlich equation (see earlier):

$$M_{ml} = (\log_e((5.06 - E_l corr) / (5.06 + 0.453))) / -0.1326 \qquad \text{(Equation 2.4)}$$

where $E_l corr$ is milk energy yield (MJ/kgW$^{0.75}$) corrected for weight loss.

$E_l corr$ is calculated as:

$$E_l corr = (E_l + E_{lwc}) / W^{0.75} \qquad \text{(Equation 2.5)}$$

Milk net energy yield, E_l, is taken as the product of milk yield (Y, kg/d) and the energy value of milk ([EV$_l$], MJ/kg) based, in the example below, on the fat concentration in milk ([FAT], g/kg) (see Tyrrell and Reid, 1965 for the full range of equations).

$$[EV_l] = Y * (1.509 + 0.0406 * [FAT]) \qquad \text{(Equation 2.6)}$$

Activity allowance (M_{act})

The allowances for standing, vertical movement and body position change are assumed to be included in the energy requirements derived from calorimetry studies and therefore included in M_{ml}. However, the term (**0.0013 *W)/k$_m$** is added to the requirement as in AFRC (1993) to describe the distance traveled component of the activity allowance. The efficiency of utilisation of ME for activity is assumed to be the same as that for maintenance (k_m), and this requires to be calculated from MEI/TDMI as described in AFRC (1993).

Total requirement

Incorporating the above, ME requirement (M^{FiM}, MJ/d) is calculated as:

$$M_{req}^{FiM} = (M_g^{FiM} + M_c + (M_{ml} * W^{.75})) + M_{act}$$

where M_g^{FiM} is the ME required for gain (MJ/d), M_c to meet the needs of pregnancy post 250 days (MJ/d), M_{ml} for maintenance and milk production (MJ/kg $W^{0.75}$), W is live weight (kg) and M_{act}, the activity allowance (MJ/d).

Calculation of ME supply

The **FiM** energy model describes the requirement for ME at the production level of feeding with dairy cows. Since feed ME concentrations in feeds are by definition measured or estimated from values derived from sheep at maintenance, a number of corrections may need to be applied to relate supply to requirement (see Figure 2.2). These include the effect of level of feeding and the influence of species (Sheep v. Cattle). The reference and predictive methods for the measurement of the ME concentration in feeds is described in Chapter 5 on feed characterisation methods.

Although the correction for level of feeding and species effects should really be on the supply side, ARC (1980) and AFRC (1993) recommended that the requirement for ME be increased by a factor of 0.018 per unit increase in feeding level above maintenance (L) to allow for the reduction in digestibility at higher intakes. However, an analysis by **FiM** of 59 treatment means in which measurements were taken with sheep at maintenance and beef or dairy cattle at feeding levels up to 4.8M showed no clear relationship between [ME_p]:[ME_m] and the level of feeding. Other factors, such as forage to concentrate ratio, also failed to improve the prediction of [ME_p]:[ME_m].

It is inevitable that level of feeding effects are confounded with species since sheep are unable to achieve the high levels of feeding obtained with dairy cows. However, an analysis of 72 comparisons of OM digestibility taken from 17 comparative studies revealed no consistent difference between

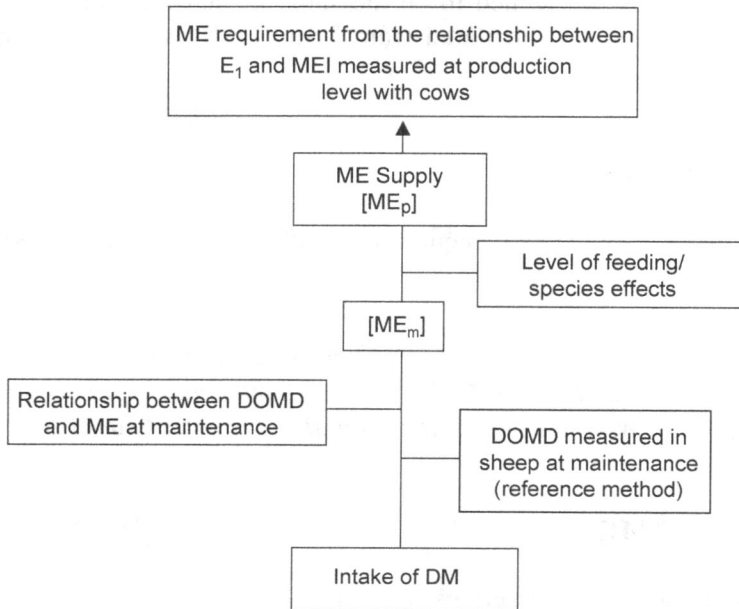

Figure 2.2 Framework for calculating ME supply

species at maintenance, particularly with mixed diets and high quality forages (Yan *et al.*, 2002). The relationship derived from these data was:

$$[ME_p] = [ME_m] * (1-0.02)$$ (Equation 2.7)

It is therefore recommended that only a small fixed factor is needed to correct ME concentration at production levels for values determined at maintenance with sheep. An independent evaluation using 13 treatment means from further experiments supported this relationship. For cows at high levels of feeding, the new correction is substantially smaller than the earlier ARC recommendation of 0.018L.

Evaluation of the FiM energy system

A series of evaluations were carried out using both calorimetric and production data. Initially, internal validation was conducted using the separate test set approach in which models were derived from a randomly selected sub-set (two thirds) of the data and tested on the remaining one third. The model performed well in these tests showing very small and non-significant bias values (-0.4 MJ/d). However, larger bias values were observed when the **FiM** energy model was tested on a number of completely independent data (see Table 2.1). Firstly, a small calorimetric data-set (42 treatment means obtained from the literature since 1976) showed a bias (**FiM** overprediction) of 7 MJ/d. Secondly, a database of 2417 treatment means from the **FiM** intake database was used as a 'production' test set. On average, **FiM** over-predicted ME requirement by 9.6 MJ/d. Investigations have so far failed to trace the source of the error. The over-prediction is in the form of a (bias) and thus is unlikely to be associated with level of feeding effects.

Table 2.1 Accuracy and precision of prediction of energy requirement

	Energy intake			*Proportion of MSPE*		
	Bias (A-P)	*s.e.*	*MPE*	*Bias*	*Line*	*Random*
Internal	-0.4	15.2	0.078	0.00	0.08	0.92
Calorimetry data	-7.0	9.8	0.07	0.33	0.07	0.60
FiM database	-9.6	22.3	0.12	0.14	0.04	0.82
FiM database adj (-10 MJ/d)	0.4	22.3	0.12	0.00	0.05	0.95

It is therefore recommended that 10 MJ of ME /d is deducted in the calculation of the allowance for ME requirement.

Conclusions

The **FiM** energy system provides (with bias adjustment) an accurate prediction using a new approach. This is fundamentally different from the previous AFRC system in that estimates of requirement are calculated empirically using relevant cows and diet types rather than built up through a factorial methodology. It provides a sound biological basis that is capable of further development and furthermore uses existing databases and feed evaluation technology.

Appendix 2.1. Data used in modelling the relationship between intake and output to derive ME requirement

	Minimum	*Maximum*	*Mean*	*s.d.*
Live weight (kg)	385	826	579	71.4
GE intake (MJ/d)	143	543	330	77.8
ME intake (MJ/d)	85	348	208	49.5
Energy outputs (MJ/d)				
Faeces	30	169	89	29.2
Urine	3	27	12	3.6
Methane	8	34	22	4.8
Heat production	68	185	126	24.3
Milk	17	160	80	28.5
Energy balance (MJ/d)	-80	84	2	22.0

Appendix 2.2 Range of functions evaluated in defining the relationship between milk energy output and ME Intake

Models	*Equations*
Linear	$E_{l(0)} = 0.59\ \text{MEI} - 0.34$
Mitscherlich	$E_{l(0)} = 5.06 - (5.06 + 0.453)*\text{EXP}(-0.1326*\text{MEI})$
Rectangular hyperbola	$E_{l(0)} = (14.3 + 0.34)\ \text{MEI}/(22.95 + \text{MEI}) - 0.34$
Gompertz	$E_{l(0)} = 0.34*\text{EXP}[1 - \text{EXP}(-0.74\ \text{MEI})*\text{LN}((4.9 + 2*0.34)/0.34)] - 2*0.34$
Logistic	$E_{l(0)} = 0.34*(1.2 + 2*0.34)/[0.34 + (1.2 + 2*0.34)*\text{EXP}(-1.67\ \text{MEI})] - 2*0.34$

3   Protein Requirement and Supply

D.I Givens, C. Rymer, B.R. Cottrill, N.W. Offer and C. Thomas

Background

In 1980, the Agricultural Research Council introduced a new approach to estimating the requirements and supply of dietary protein for ruminant livestock (ARC 1980). This system, which was based on a framework proposed by Burroughs *et al.* (1975), introduced the principle of metabolisable protein (MP), in which feeds were characterised in terms of the extent to which they were degraded in the rumen to provide nitrogen (N) for microbial protein synthesis. Microbial protein, together with dietary protein not degraded in the rumen provided the basis for estimating the supply of protein available for metabolism by the host animal.

At about the same time, a number of other factorial protein models were published in Europe, North America and Australia. Together with the MP system, they shared a common framework and adopted a common nutritional currency (metabolisable protein), but differed in terminology and the extent to which they were dynamic. They included the PDI (Vérité *et al.*, 1987), AAT/PBV (Madsen, 1985), AP (NRC, 1985,), DVE (Tamminga *et al.*, 1994), the Australian (CSIRO, 1990) and the Cornell Net Carbohydrate and Protein System (CNCPS) (Russell *et al.*, 1992; Sniffen *et al.*, 1992).

Protein requirements

Background and principles

The range of systems currently available to calculate the supply and requirement for protein has been outlined above. Although the MP system (ARC, 1980; AFRC, 1993) shares a common framework with these, it differs in a number of ways in respect to the calculation of requirements. In particular, estimates of MP for maintenance (MP_m) are significantly lower in the UK MP system than in any of the others currently in use. For example, for a 600 kg cow consuming 22 kg DM/d, with no pregnancy or liveweight change MP_m (g/d) ranges from 394 (PDI) to 774 (NRC)[1]. This compares with an estimate from the UK MP system of 278g/day.

In the MP system as described in AFRC (1993), estimates of requirements were based on basal endogenous N losses, but *at a maintenance level of feeding*, and no adjustment was made to account for losses at higher intakes. As a result, the UK system would in relative terms underestimate requirements or over-predict production. This was confirmed by published comparisons with other European systems (PDI, AAT/PBV and DVE), which suggested that the MP system had the highest mean prediction error (van Straalen *et al.*, 1994; Tuori *et al.*, 1998). These authors concluded that this was due to the low maintenance requirement in the MP system relative to other models, together with the fixed - and high - value for the efficiency of utilization of MP for milk protein synthesis (k_{nl}).

A new model to calculate MP_m should therefore include an estimate of metabolic faecal N loss related to dry matter intake. The NRC (2001) is the most recent of the 'new' models that include such a correction for intake and because of the substantial database from which it was derived it is recommended that the NRC (2001) coefficients for calculating maintenance requirements are used for MP^{FiM}.

In common with other published systems, the MP system is particularly sensitive to variations in the efficiency of utilization of MP for milk protein synthesis (k_{nl}). Efficiency ranges from 0.64 (PDI) to 0.73 (AAT/PBV), while in the DVE system it is a variable, decreasing with increasing milk yield. Prior to the commencement of *Feed into Milk*, four major feeding studies were undertaken at ADAS (the last in conjunction with SAC), to test various components of the MP system (Newbold *et al.*, 1994). It was concluded that in circumstances in which MP supply was clearly deficient relative to ME, then the value for k_{nl} proposed by the MP system (0.68) was appropriate. In situations where protein is in excess, k_{nl} was lower, by between 8% and 14% than that currently recommended. Because MP^{FiM} is designed to meet requirement for both energy and protein, it is proposed that the value for k_{nl} of 0.68 should be retained.

---

[1] In the DVE system it is 114 g/ d. However, metabolic nitrogen losses are attributed to the feed rather than to the animal.

Calculation of MP^FiM requirements

Requirements for maintenance and endogenous loss

Cows at maintenance continue to lose protein from their bodies in urine (endogenous urinary protein), in hair and scurf and through losses into the digestive tract (secretions, enzymes, sloughed cells, etc).

MP^{FiM} uses NRC (2001) coefficients for estimating MP requirements (g/d) for endogenous urinary protein and hair and scurf protein:

- Endogenous urinary protein $4.1*W^{0.5}$

- Hair and scurf $0.3*W^{0.6}$

Much of the endogenous N entering the digestive tract is re-absorbed, either directly or after degradation into ammonia and incorporation into microbial protein. That portion that is excreted can be measured as metabolic faecal protein (MFP, g/d). NRC (2001) recognises that losses increase with intake and describe this as:

$MFP = 30DMI$,

where DMI is the total dry matter intake in kg/d.

Losses of endogenous N to the hind gut also appear in faeces, much of it in the form of bacteria synthesized in the caecum and hind gut. MP^{FiM} adopts the approach of NRC (2001) by including an adjustment for that fraction of intestinally indigestible rumen-synthesized microbial protein that is degraded (and absorbed as ammonia) from the hind gut. This is estimated as:

$0.5((DMTP/0.8)-DMTP)$

where DMTP is digestible microbial true protein (g/d).

In addition, there is an endogenous protein (EP) correction. This is based on artificial feeding experiments (Ørskov *et al.*, 1986) and some N^{15} studies. EP supply at the intestine is estimated to be about 15% of NAN flow. The EP is sloughed cells, secretions, enzymes etc. However, NRC (2001) also add EP to the supply side for and then add the same amount divided by 0.67 to the maintenance requirement side (assumes that efficiency of synthesis of EP from MP is 0.67). The net effect on MP balance is 2.34DMI. The calculation is shown in Appendix 3.1.

Calculation of MP_m requirements for maintenance (g/d) in the MP^{FiM} model is therefore:

$$MP_m{}^{FiM} = 4.1W^{0.5} + 0.3W^{0.6} + 30TDMI - 0.5((DMTP/0.8)-DMTP) + 2.34DMI$$

(Equation 3.1)

where W is liveweight (kg) DMI is total dry matter intake (kg/ day) and DMTP is digestible microbial true protein (g/d)

Requirement for preganancy, milk and body weight change

The estimates of MP requirement for pregnancy (MP_c), milk (MP_l), and body weight gain and loss (MP_g, MP_{loss}) are as described in AFRC (1993) and the detail of these is given in the list of equations on the CD.

MP^{FiM} requirement

The total MP required (g/d) is therefore calculated as:

$$MP_{req}{}^{FiM} = MP_m + MP_c + MP_l + MP_g + MP_{loss}$$

Protein supply

Background and principles

The MP system continues to provide a sound base to predict the supply of protein to dairy cows. Furthermore it is supported by an extensive database that describes the degradation characteristics of feeds commonly used in the UK. However, the system (AFRC, 1993) has a number of weaknesses:

- Fermentable metabolisable energy (FME) as an estimate of energy supply to the rumen micro-organisms is unsatisfactory in that:

  - It is based on an often imprecise estimate of ME

  - It includes undegradable carbohydrates and proteins that do not provide energy in the rumen.

- The supply of FME is not defined dynamically so that the interaction of rates of protein degradation, rumen outflow rate and feed fermentation is not addressed.

- The yield of FME for microbial synthesis takes no account of the nature of the feed substrates fermented although this has been shown to affect microbial yield (Archimède *et al.*, 1997).

It is recommended that supply continues to be described as MP in current UK advisory practice but to address the deficiencies the following modifications are needed:

- Improved characterisation of feed energy available for microbial use.

- A dynamic description of the energy that microbes derive during feed degradation and fermentation.

- Similar mathematic descriptions of the dynamics of both energy and nitrogen degradation in the rumen.

Framework for the supply of MP^{FiM}

MP^{FiM} differs radically from the current system in that it defines energy supply for microbial protein synthesis as adenosine triphosphate (ATP) rather than FME and assumes that ATP is produced in a two-stage process (Beever, 1993).

- The microbial degradation of feeds to simple compounds.

- The fermentation of these compounds to yield ATP for microbial maintenance and synthesis.

The supply of ATP from a feed or diet (mol/kg DM) can thus be calculated from:

- The degradation characteristics of feeds measured *in situ.*

- Modified by rumen outflow to estimate the effective degradability of the DM.

- Multiplied by the yield of ATP from the effectively degraded DM (ATPy).

The potential Microbial Crude Protein (MCP) supply is then calculated from:

- The yield of ATP and

- The efficiency of microbial protein synthesis per unit of ATP (Y_{atp})

There is ample evidence that the extent of feed degradation is related both to feed characteristics and rumen outflow rate. Also, the efficiency of ATP production per unit of degraded feed can vary considerably due mainly to feed type. Furthermore it is known that the efficiency of microbial protein synthesis largely depends on outflow rate (Hespell, 1979).

Thus, the rate and extent of the transactions at each stage of the process described above is dependant on the composition of the diet and the rumen outflow rate. The framework for energy supply within the rumen now corresponds with the principles and mathematics set out for protein degradation as described in AFRC (1993). An outline of the framework is presented in Figure 3.1.

Degradation of DM and Nitrogen

Degradability characteristics of feeds are estimated from parameters derived from a modified *in situ* technique. This is recommended as the **FiM** reference method in preference to *in vitro* procedures:

- To address concerns that *in vitro* techniques involving the use of dried, milled feeds can lead to erroneous estimates particularly for forages and relatively unprocessed feeds (e.g. Sanderson *et al.*, 1997; Givens and Gill, 1998).

- To recognise that an extensive database of *in situ* measurements for feeds already exists.

The standard *in situ* method involving the suspension of polyester bags in the rumen (Ørskov and Mehrez, 1977) provides data on the proportional degradation of feed components to which an exponential function is fitted of the form

$$dg = a + b\{1 - e^{(-ct)}\}$$

where a = *immediately soluble component*
b = *potentially degradable component, other than 'a' and*
c = *fractional rate of degradation of the b component per hour.*

It is important to recognise that the 'a' value is in reality the washable rather than soluble component. Also the technique is deficient in that any fine particles lost through the bag pores are treated as if they were part of the immediately soluble feed pool (Dhanoa *et al.*, 1999). This can be overcome if the technique is enhanced with estimates of water solubility and thus giving a solubility (s) value for the DM and N for each feed (Hvelplund and Weisbjerg,

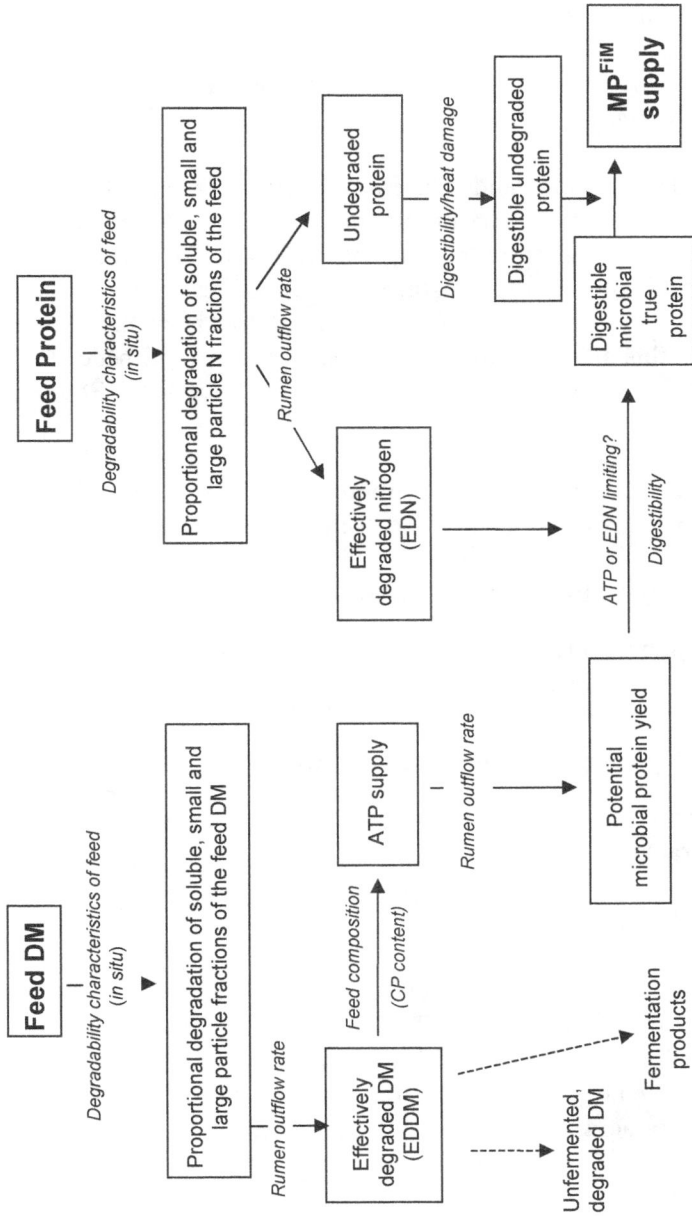

Figure 3.1 Framework for the supply of MP^FiM

2000). Fine particles are then calculated as the difference between the *in situ* initial wash value ('a') and s but importantly, the fine particles are assumed to flow out of the rumen with the liquid phase.

Outflow rates and the estimation of the effective degradabilities of DM (eddm) and N (edn)

In order to derive eddm and edn from *in situ* values, estimates of rumen outflow rate are needed. The prediction of outflow rate in AFRC (1992) is based solely on plane of nutrition, and makes no differentiation between liquid and solid phases of digesta, or between forage and concentrate. It also makes no allowance for the proportion of forage in the diet, although it is known that this affects outflow rate (Sauvant and Archimède, 1989). These factors are addressed in the models of Owens and Goetsch (1986) and Sauvant and Archimède (1989).

The model of Sauvant and Archimède (1989) is adopted for MP[FiM] in preference to that of Owens and Goetsch (1986) because:

- The database from which it was developed is larger.

- The database includes more European diets.

- It uses metabolic body size rather than liveweight as a predictor.

- It has higher coefficients of regression.

The model provides separate prediction of the outflow rate of liquids, forages and concentrates (Figure 3.2).

The yield of ATP

An *in vitro* procedure is recommended as the reference method to estimate the efficiency of production of ATP per unit of degraded DM. In summary, feeds are incubated with buffered rumen fluid and the yield of ATP (moles/g degraded DM, ATPy) is then calculated from the measured production of short chain fatty acids (SCFA) and DM degraded. The details of the method, along with prediction equations, are presented in Chapter 5 on Feed Characterisation methods.

The concept of efficiency of utilisation of ATP for microbial dry matter production (Y_{ATP}) is retained in the MP[FiM] system but with the important

Feed DM

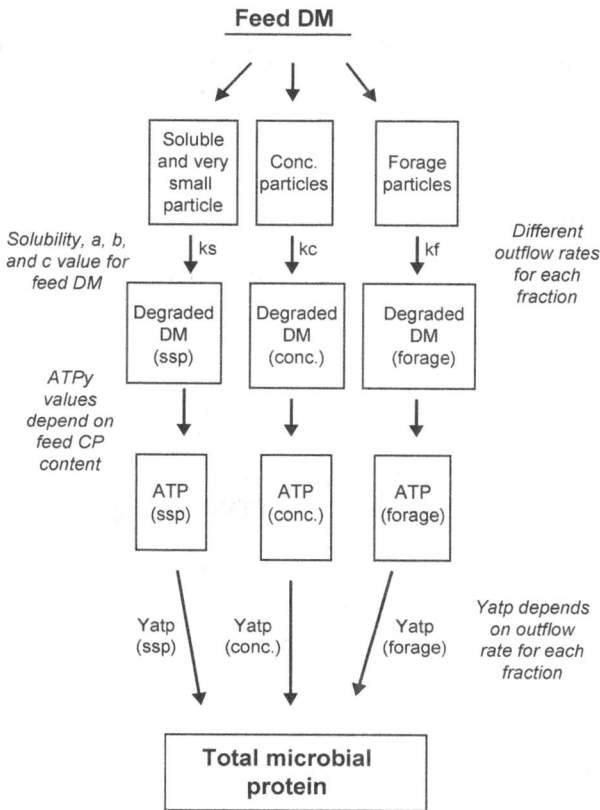

Figure 3.2 Framework for potential microbial protein production from feed dry matter

variation that Y_{ATP} is now a function of fractional outflow rate. Thus three Y_{ATP} values for the liquid, forage and concentrate fractions are defined and the amount of microbial DM (MDM) is calculated for each of the three fractions from the product of ATP supply and the efficiency of utilisation of ATP for microbial synthesis. In turn MDM is converted to microbial crude protein (MCP^{FIM}) supply. Finally, the test as to whether the estimated energy supply (now described as ATP) or the supply of nitrogen (EDN) is limiting MCP production remains central to the system as in AFRC (1993).

Post-rumen supply of MP is calculated as the sum of digestible undegraded dietary protein and digestible microbial protein, as in previous versions of the MP system (AFRC, 1993).

Calculation of MP^FiM supply

The framework for the system is shown in Figure 3.1. The process to determine the actual supply of microbial crude protein (MCP) is:

For each feed in the ration calculate:

- The potential microbial CP arising from ATP (MCP_{atp}).

- The potential microbial CP arising from EDN (MCP_{edn}).

Sum these values for the total diet:

- If MCP_{atp} is equal to or less than MCP_{edn} then the actual MCP supply from the total diet will be limited to MCP_{atp} (i.e. ATP limiting).

- If MCP_{atp} is less than MCP_{edn} then the actual MCP supply from the total diet will be limited to (MCP_{edn}) (i.e. EDN limiting).

In this section the equations to determine the fractional outflow rates and the potential production of microbial protein from either ATP or EDN will be described. A full list of the equations can be found on the CD.

Fractional outflow rates

The supply of ATP and its utilisation by rumen microbes and also the supply of effectively degraded nitrogen, are all crucially dependant on the outflow rate of material from the rumen (k, proportion/h), This is calculated for three pools from the equations of Sauvant and Archimède, (1989):

$$k_{liq} = 0.0245+(0.25 \ DMI/(W^{0.75}))+0.04f^2 \qquad \text{(Equation 3.2)}$$

$$k_f = 0.0035 + (0.22 \ DMI/(W^{0.75}) +0.02f^2 \qquad \text{(Equation 3.3)}$$

$$k_c = 0.0025 + 0.0125 \ k_f \qquad \text{(Equation 3.4)}$$

where k_{liq}, k_f and k_c are the fractional outflow rates (proportion/h) of liquids, forages and concentrates respectively, DMI is dry matter intake (kg/d), W is liveweight (kg) and f is the proportion of forage in the diet (DM basis).

ATP supply and its utilisation for the potential supply of MCP

To derive the ATP supply from a feed, firstly the effective degradability of the following fractions need to be calculated using the characteristics derived from the modified *in situ* technique.

- The soluble and small particles (SSP).

- Large particles (LP) from either concentrates or forages.

The effective degradability of the soluble and small particle fraction of the feed (eddm$_{ssp}$) is derived from the following equation if it assumed that the rate of degradation of these fractions is 0.9/h:

$$eddm_{ssp}=(0.9s/(0.9+k_{liq}))+(\beta_D c/(c+k_{liq}))$$ (Equation 3.5)

where s is the soluble DM proportion; k_{liq}, the fractional outflow rate of liquid; β_D, the degradable small particle DM proportion of the feed; c, the fractional rate of DM degradation of the b fraction and b, the degradable large particle DM proportion of the feed (commonly referred to as the potentially degradable DM fraction).

For silages the s fraction is first corrected to account for fermentation acids ([TFA], g/kgDM) that do not yield ATP for microbial synthesis so that:

$$s_{con} = s - ([TFA]/1000)$$

The degradable small particle DM content of the feed (b$_D$) content of the feed is calculated from the equation:

$$\beta_D=(b(a-s))/(1-a)$$ (Equation 3.6)

where a is the in situ initial loss value calculated as the intercept of the curve obtained when time is plotted against DM degradability and s the soluble DM fraction.

The eddm of the large particle fraction of either forage or concentrates (eddm$_{lp}$) is defined as:

$$eddm_{lp}=(bc/(c+k))$$ (Equation 3.7)

where k is the fractional outflow rate for the particular pool of material as k_f or k_c for LP, depending on whether the feed is a forage or a concentrate).

The ATP supply (mol/d) that comes from the feed's SSP and LP fractions is calculated from:

$$ATP_{ssp} =(eddm_{ssp} * DMI * ATPy)$$ (Equation 3.8)

and

$$ATP_{lp} = (eddm_{lp} * DMI * ATPy)$$ (Equation 3.9)

where ATP_{ssp}, ATP_{lp} is the supply of ATP from the SSP and LP fractions of the feed respectively, DMI is the DM intake of the feed (kg/d) and ATPy the yield of ATP (mol per kg of DM degraded).

ATPy may be measured directly or predicted from the equation:

$$ATPy = 27.34 - 0.0248[CP]$$

where [CP] is the crude protein content of the feed (g/kg DM) (see Chapter 5 and note that this equation may not be appropriate for feeds high in Non Protein Nitrogen):

The amount of microbial dry matter (MDM, g/d) that may then be produced from this ATP supply is calculated from:

$$MDM = (ATP_{ssp} \times Y_{ATP\ ssp}) + (ATP_{lp} \times Y_{ATP\ lp})$$ (Equation 3.10)

where $Y_{ATP\ ssp}$ and $Y_{ATP\ lp}$ are the efficiencies of MDM synthesis (g microbial dry matter/mol ATP) from the SSP and LP fractions respectively.

The value of Y_{ATP} is calculated separately for each of the three pools from the equation:

$$Y_{ATP} = 9 + 50k$$ (Equation 3.11)

where k is the fractional outflow rate for the particular pool of material (k_{liq} for SSP and k_f or k_c for LP, depending on whether the feed is a forage or a concentrate).

To convert MDM to microbial crude protein, it is assumed that the N content of MDM is 100 g N/kg DM, and that microbial crude protein consists of 160 g N/kg crude protein. The equation for converting MDM to the supply of MCP[FIM] (microbial crude protein supply from feed, g/d) is therefore:

$$MCP_{atp} = MDM \times 0.1 \times 6.25$$ (Equation 3.12)

This value represents the amount of MCP that may be produced from ATP if there is sufficient EDN available in the rumen for its synthesis.

Supply of effective degradable N from each feed

The effective degradability of the N fraction (edn) of each feed is calculated from the degradability characteristics of the N fractions determined *in situ* and the appropriate outflow rates for the three pools using the equation:

$$edn=(0.9s_N/(0.9+k_{liq}))+(b_{DN} c_N/(c_N+k_{liq}))+(b_N c_N/(c_N+k)) \quad \text{(Equation 3.13)}$$

where s_N, b_{DN}, c_N and b_N are respectively the soluble, the degradable small particle, the fractional rate of degradation and degradable large particle N fractions of the feed.

As with eddm, k is the fractional outflow rate of the large N particles of the feed, and either k_f or k_c is used, depending on whether the feed is a forage or a concentrate.

The value of β_{DN} (degradable small particle N proportion) in the above equation is:

$$\beta_{DN}=(b_N(a_N-s_N))/(1-a_N) \quad \text{(Equation 3.14)}$$

where a_N is the intercept of the curve obtained when time is plotted against N degradability.

The supply of effectively degraded N (g/d) from each feed is calculated from:

$$EDN \text{ supply} = edn * DMI *[N] \quad \text{(Equation 3.15)}$$

where DMI is the dry matter intake of the feed (kg/d), and [N] the nitrogen content of the feed (g/kg DM).

Assuming the N content of protein to be 160 g N/kg crude protein, the potential MCP^{FIM} that may be synthesised from EDN supply from a feed is:

$$MCP_{edn} = EDN * 6.25 \quad \text{(Equation 3.16)}$$

Undegradable protein

The other component of the metabolisable protein supply (MP^{FIM}) is the supply of digestible undegraded protein (DUP^{FIM}) from each feed. Firstly the supply of undegraded protein (UDP) for each feed is calculated as:

$$UDP = ([CP] * DMI) - (EDN * 6.25)$$ (Equation 3.17)

where [CP] is the crude protein content of the feed (g/kg), DMI is the DM intake of the feed (kg/d) and EDN the effective degraded N supply from the feed.

As in the MP system, it has been assumed that the protein associated with the ADIN fraction is not digestible and that the digestibility coefficient of the remainder of the rumen undegradable protein is 0.9.

$$DUP^{FIM} = (0.9 * UDP) - (DMI * [ADIN] * 6.25)$$ (Equation 3.18)

where [ADIN] is the concentration (g/kgDM) of acid detergent insoluble N in the feed and DMI is the DM intake of the feed (kg/d).

Total diet MP supply

The potential yields of MCP for either ATP or EDN limiting situations (MCP$_{atp}$ and MCP$_{edn}$) are derived for the total diet by summing the respective values from the individual feeds. The total diet supply of MCP^{FIM} becomes the lower of the two estimates.

When the actual MCP^{FIM} supply from the total diet has been derived, it must be converted into digestible microbial true protein (DMTP^{FIM}). As in the MP system (AFRC, 1993) it is assumed that the true protein content of MCP is 750 g /kg CP, and that the digestibility of the true protein is 850 g/kg. The DMTP^{FIM} supply is therefore:

$$DMTP^{FIM} = 0.75 * 0.85 * MCP^{FIM} = 0.6375 * MCP^{FIM}$$ (Equation 3.19)

The DUP supply for the diet is the sum of the DUP supplies from each individual feed.

The supply of MP^{FIM} is then calculated from:

$$MP^{FIM} = DMTP^{FIM} + DUP^{FIM}$$ (Equation 3.20)

Evaluation of MP^{FiM}

Given that it is not possible to evaluate the requirement component separately from assumptions of the supply model, the approach was firstly to test the components related to the output from the rumen model against *in vivo* data and then to examine the accuracy of the whole system using production data.

Microbial efficiency

Archimède *et al.* (1997) reviewed 320 *in vivo* observations of the efficiency of microbial N synthesis (EMNS, g microbial N/kg OM truly degraded in the rumen). The mean EMNS observed was 23.5 g microbial N/kg OM truly degraded in the rumen (OMTDR) but the nature of the carbohydrate present in the diet had a substantial bearing on the values with the highest values being seen for starch-rich diets. The mean and the range of values reported by Archimède *et al.* (1997) were compared with those used in the current MP system and those measured on 39 feeds according to the MP^{FiM} approach assuming a Y_{ATP} of 12. The results are shown in Table 3.1.

Table 3.1. Comparison between estimates of the efficiency of microbial protein supply (EMNS)

| | *EMNS (g microbial N/kg OMTDR)* | | |
	*In vivo (n=320)**	*MP***	*FiM*
Mean	23.5	26.7	23.5
Minimum	14.2	24.0	16.1
Maximum	32.8	29.4	28.5

*Archimède *et al.* (1997)
** In the MP system, EMNS changes with level of feeding, but supply of FME does not. The mean value presented is for a growing animal (outflow rate 0.06/h), while the minimum and maximum is calculated for outflow rates of 0.02/h (maintenance) and 0.08/h (lactating cow).

The estimates by the **FiM** approach described the range of EMNS values observed *in vivo* considerably better than did the current MP system and in accord with Archimède *et al.* (1997). The highest values were for starch or digestible fibre-rich feeds. Thus the **FiM** approach is able to discriminate between feeds (and ultimately diets) in a way that FME cannot.

The range of values of Archimède *et al.* (1997) also agree well with the mean and range of values given in NRC (2001) for diets where the rumen N balance was zero or positive. In a further evaluation, microbial N flow was predicted using the MP or **FiM** model and compared with the relationship derived from *in vivo* values by NRC (2001). It can be seen from Figure 3.3 that **FiM** provides much more reliable estimates of microbial N yield than does the previous MP model. The MP system overestimates microbial N yield, particularly at high levels of feed intake and hence milk production.

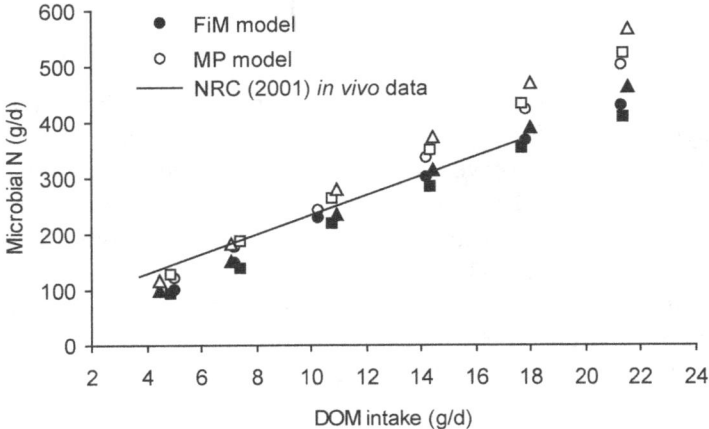

Figure 3.3 Relationship between **FiM** and MP estimates of microbial N supply, and *in vivo* observations.

Evaluation of the whole MP^FiM system

Data from five dairy cow studies were used to evaluate the **FiM** protein model. The first four experiments were carried out at ADAS Bridgets and SAC (Newbold *et al.*, 1994; Metcalf *et al.*, 1997). Each study was designed to test different aspects of the MP system (AFRC 1993). The first three involved 21 different diets, varying in ERDP:FME ratio, protein degradability and MP supply. In the fourth, the diets supplied either 25% or 12.5% above, or 25% or 12.5% below that recommended by the MP system, while ME was supplied at fixed levels to meet requirements. An experiment undertaken by SAC, which was also part of this series of experiments, examined more closely the relationship between MP supply and ME. These studies provided data from 24 different dietary treatments. The ranges in intake and milk production are shown in Appendix 3.2. A further study at ARINI, measured responses to different amounts of concentrates of varying protein content and quality (Mayne, 1990; Mayne, *unpublished*).

Calculations were based on treatment means for groups of cows and for a total of 50 diets. The MP^{FiM} model was used to calculate MP^{FiM} supply corresponding to each diet. Requirements were calculated using observed cow characteristics and milk output. Where possible, measured values were used for feeds although, in most cases, the degradation characteristics of the feed ingredients were obtained from the MP^{FiM} database.

Figure 3.4 shows a plot of MP^{FiM} supply (MPs) against milk true protein (TP) output for the test data set. Diets from the five experiments were divided into three groups on the basis of the MP^{FiM} balance calculated using the MP^{FIM} model.

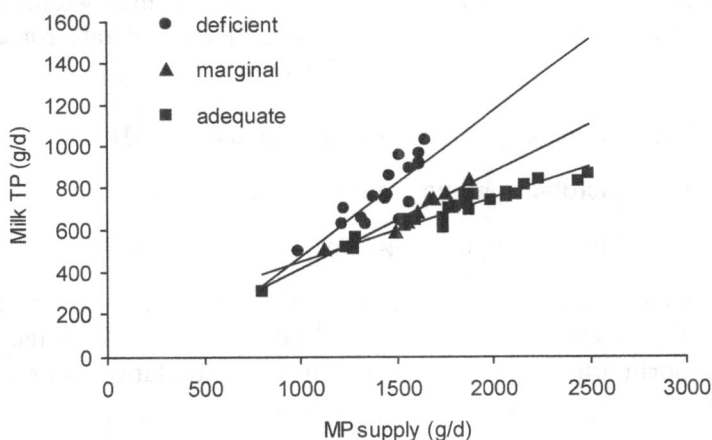

Figure 3.4 Milk true protein response to MP supply

- *Group 1.* *Deficient diets* providing <=98% of MP^{FiM} requirement

  Milk TP (g/d) = 0.70 MPs –233 R^2 = 0.71

- *Group 2.* *Marginal diets* providing >=98 and <=102% of MP^{FiM} requirement

  Milk TP (g/d) = 0.46 MPs –45 R^2 = 0.91

- *Group 3.* *Adequate diets* providing >102% of MP^{FiM} requirement

  Milk TP (g/d) = 0.30 MPs +139 R^2 = 0.92

The evaluation suggests that the **FiM** protein model performs as a 'requirement based' model should. Unlike evaluations with the past MP system (van Straalen *et al.*, 1994), it correctly distinguishes diets showing big responses to additional MP (*MP^FiM deficient*) from those that show only small responses (*MP^FiM adequate*). The slope of milk TP output/MP^FiM supply for deficient diets (0.70) is close to published values for maximum efficiency, and justifies the adoption of $k_{nl} = 0.68$. Furthermore, the MP^FiM requirements for maintenance, estimated by extrapolation, are close to values predicted by the model.

Use in practice for linear programming of feed compounds

For the formulation of compound dairy feeds using linear programming techniques, fixed estimates of MP supply are required. Given that the variation in outflow rates is considerably greater between rather than within the liquid, forage and concentrate pools, an assumption of a fixed rate for each pool allows the calculation for each feed of 3 values.

> MP from microbial protein – energy limited   (**MPE**)
>
> MP from microbial protein – nitrogen limited   (**MPN**)
>
> MP from digestible bypass protein   (**MPB**)

Such values can assist in the formulation of diets to optimise the supply of ATP and EDN to rumen microorganisms. In these circumstances the following values (proportion/h) are suggested based on a simulation over a range of concentrate to forage ratios (see Appendix 3.3):

$$k_{liq} = 0.08$$

$$k_f = 0.045$$

$$k_c = 0.06$$

Conclusions

The MP^FiM system to predict the requirement and supply of protein provides:

- An improved dynamic description of the energy that microbes derive from degradation of feeds.

- A supply of energy to microbes quantified in terms of ATP.

- A partition of feeds into 3 fractions of feed DM which provides a base for further modification as new information emerges.

- A diet dependant estimate of yield of ATP per unit of degraded DM.

- A dynamic description of the efficiency of use of ATP for microbial N synthesis.

- Predictions of microbial efficiency and yields that agree well with *in vivo* estimates.

- A 'requirement-based' model that performs as it should in that it correctly distinguishes diets showing big responses to additional MP (*MP^{FiM} deficient*) from those that show only small responses (*MP^{FiM} adequate*).

The feed characterisation methodology for the system, the use of the model in practice together with a support system that gives guidance on amino acid requirements is presented in Chapter 5 and on the CD.

Appendix 3.1 Correction for Endogenous Protein

NRC (2001) add EP to the supply side for **MP = 0.4*11.875*DMI** and then add the same amount divided by 0.67 to the maintenance requirement side (assumes that efficiency of synthesis of EP from MP is 0.67). The net effect on MP balance is:

$$0.4*11.875*DMI - (0.4*11.875*DMI/0.67)$$

$$= -0.493*0.4*11.875*DMI$$

$$= -2.34*DMI$$

Appendix 3.2. Ranges in intake and milk production from the ADAS Bridgets and SAC MP studies used to test the whole system.

	Mean	*Min*	*Max*	*SD*
Liveweight (kg)	607	566	654	27.3
Silage intake (kg DM/d)	11	10.2	11.7	0.41
Compound intake (kg DM/d)	8	6.8	8.6	0.81
Total DMI (kg/d)	18.4	17	20.1	0.87
Milk yield (kg/d)	26.8	20.7	34.4	3.41
Milk protein (g/kg)	29.93	27.6	32.3	1.12
Milk fat (g/kg)	42.10	36.2	45.6	2.33

Appendix 3.3 Effect of practical feeding regimes on outflow rates.

Forage/Concentrate ratio	*78/22*	*54/46*	*46/54*	*36/64*	
Milk yield (kg/d)	*20*	*30*	*40*	*50*	*Mean*
kliquid	0.081	0.075	0.077	0.079	0.078
kforage	0.044	0.043	0.047	0.049	0.045
kconc	0.058	0.056	0.061	0.064	0.060

Frank Wright Ltd (personal communication)

4   Decision Support Systems to aid the Process of Ration Formulation

N.W. Offer, D.I. Givens, J.S. Blake and C. Thomas

A set of equations that allow rations to be formulated that meet requirements for protein and energy is only the beginning of the process to provide the cow with nutrients that meet her needs. Because such systems only deal with energy and protein as currencies, value judgements have to be made to ensure for example that the diets formulated promote a stable rumen environment, provide an optimum balance of amino acids and ensure that milk composition is not compromised. This section describes the principles and calculations for three decision support systems (DSS) that can be used to aid the ration formulation process.

This process of ration formulation is dealt with in more detail on the accompanying CD where there are also examples of the use of the three DSS.

A decision support system to aid the maintenance of rumen stability

Background

The **FiM** rumen model that predicts the supply of ATP and the synthesis of microbial protein assumes that rumen conditions are stable and, in particular, that the pH of rumen contents is close to the 'optimum' value of 6.2. To predict the effect of deviations from this value on rumen metabolism and the end products of fermentation and digestion would require a complex mechanistic model beyond the scope of **FiM** (see AFRC TCORN Report No 11, 1998). Nevertheless, a system to ensure that diets formulated do not cause instability should be central to any rationing system and in the context of **FiM** to the correct operation of the rumen model described on the CD.

Principles and methods

One of the principle means by which rumen pH is regulated is by the salivation of the cow, which is stimulated by chewing. Some means of predicting how much chewing would result from the formulation of a particular diet (or by the selection of a particular feed) is therefore required. A system to ensure rumen stability also requires a measurement of the acid load that a feed (or diet) puts on the rumen. This is a measure of the acid content of the feed and the likely amount of acid produced by the rumen fermentation of that feed. However, this acid load is balanced by the inherent buffering capacity of the feed that will assist in the regulation of rumen pH (McBurney *et al.*, 1983). Measuring the 'acidogenicity', or the potential acid load (PAL) of the feed takes into account both of these factors.

The Structural Value System (SVS) developed by de Brabander *et al.* (1996) focuses on the ability of feeds to buffer rumen pH through the promotion of chewing and saliva production. It does not make allowances for variation in intrinsic feed buffering capacity or of acids contained in feeds. For example, extensively fermented, highly acidic silages are considered to have equal effects on rumen pH as wilted silages of restricted fermentation. The SVS has been tuned by de Brabander *et al.* (1996) to provide 'sensible' answers and the derivation of some of the relationships is not clear. However, it is considered to provide a sound practical basis for the required DSS and can be improved by the inclusion of PAL.

Accordingly, it is recommended that RSV (Rumen Stability Value) replaces the term 'Structural Value' because RSV includes both acid buffering effects (Structural Value) and the acidogenicity of the diet (its potential acid load or PAL).

The requirement for RSV is affected by:

• the age of the cow,

• milk yield and quality, and

• the feeding system that is used.
 (de Brabander *et al.*, 1996)

The supply of RSV by a feed is a function of its:

• fibre content (as estimated by neutral detergent fibre, NDF)

• feed type (forage or concentrate)

• potential acid load (PAL).

The NDF content provides an estimate of the amount of chewing activity that will be stimulated by eating the feed, and the salivation produced during chewing provides one of the means by which rumen pH is controlled by the cow. The system recognises that NDF from forage has a greater effect on stabilising rumen pH than NDF from concentrates because of the differences in the physical form of the two sources.

The PAL content is a measure of the acid load in the rumen and hence provides an estimate of the amount of buffering that needs to be done to maintain a stable rumen pH. It should take account of:

• the amount of acid already in the feed (eg with ensiled forages),

• the amount of acid that will be produced by the rumen fermentation of the feed, and

• the inherent buffering capacity of the feed (which will reduce the requirement for buffering activity by the cow).

Calculation of the Requirement and Supply of RSV

Requirement for RSV

The de Brabander SVS system is used to calculate the requirement for RSV (standard requirement = 100):

$$RSV_{req} = 100 + yfac + fatfac + lacfac + mealfac \qquad \text{(Equation 4.1)}$$

where yfac (milk yield factor) = $100 * ((y - 25) * .01)$
fatfac (milk fat concentration factor) = $100 * ((44 - fat) * .005)$
and y is milk yield (kg/d)and [fat] the milk fat concentration (g/kg)

lacfac (lactation number factor) = 0 for lactation numbers <4
= -7 for fourth lactation
= -15 for fifth and subsequent lactations
mealfac (meal frequency factor) = 10 for 1 concentrate meal per day
= 5 for 2 concentrate meals per day
= 0 for 3 or 4 concentrate meals per day
= -5 for 5 concentrate meals per day
= -10 for 6 or more concentrate meals per day or for TMR

RSV supply

The RSV of a feed is calculated separately for forages and concentrates using the equations of de Brabander *et al.* (1996) modified for the effect of the potential acid load (PAL, meq/kgDM). It is assumed that feeds with a PAL of 800 (e.g. hay) have a neutral effect on rumen pH in terms of their buffering capacity and fermentation acid production. Thus feeds with higher PAL values tend to lower rumen pH and need more NDF to balance this effect.

Forages: $RSV = 100 * (0.006 * [NDF] - 0.001 * ([PAL] - 800))$

(Equation 4.2)

Concentrates: $RSV = 100 * (0.175 + 0.00082 * [NDF] - 0.001 * ([PAL] - 800))$

(Equation 4.3)

where [NDF] is the concentration of neutral detergent fibre (g/kgDM) and [PAL], the potential acid load (meq/kgDM).

PAL is determined *in vitro* by incubating feeds with rumen liquor under standard conditions and measuring the total free acid produced. The reference method uses a sensitive electrometric titration of the mixture following incubation. This produces an absolute estimate of PAL in meq/kg DM feed. A NIRS prediction method was developed for grass silage and an equation is available to predict PAL in maize silages. (See Chapter 5 on feed characterisation methods).

The total RSV supply from a diet is calculated as the sum RSV for each feed multiplied by its intake (DMI):

$RSV = RSV_{feed} * DMI_{feed}$

(Equation 4.4)

RSV balance

RSV balance is then derived to compare supply (as a concentration in the total diet, [RSV]) with requirement (RSVreq):

$[RSV] = RSV / TDMI$

(Equation 4.5)

$RSVbalance = [RSV] - RSV_{req}$

(Equation 4.6)

The PAL and RSV values for a range of feeds are shown below in Table 4.1

Table 4.1. **Example PAL, NDF and RSV contents of a range of feeds.**

Feed	PAL content (meq/kg DM)	NDF content (g/kg DM)	[RSV]
Hay	800	600	360
Wilted grass silage	900	480	278
Fermented grass silage	1100	480	258
Barley	1150	211	0
Molassed sugarbeet feed	900	321	34

The interpretation of RSV balances

As in the system of de Brabander *et al.* (1996), if the balance of RSV supply is greater than 20, then it is assumed that no problems with rumen acidosis should be encountered when that diet is offered to the cows for which it was formulated. The risk of rumen acidosis increases as the RSV balance decreases. The model therefore provides a set of warnings as the balance falls below 20, and if it is less than 0, it is strongly recommended that the diet be modified to increase the RSV.

Detail of the interpretation of RSV together with examples of diets formulated using the RSV system is provided on the accompanying CD.

A decision support system to predict the supply and adequacy of amino acids

Background and principles

A number of studies have confirmed that the amino acid make-up of metabolisable protein (MP) can vary substantially and may not always be optimum for milk protein synthesis. This, together with the wide range in diets used in the feeding of dairy cows requires the consideration of the adequacy of the supply of, and requirement for, individual amino acids. Accordingly, a decision support system was derived which calculates the supply and judges the adequacy of a range of amino acids (AA). In summary, the methodology assumes:

- a fixed AA composition of microbial protein

- no preferential rumen degradation of AA within a feed

- no preferential absorption of any AA in the small intestine

- a set of threshold values based on *in vivo* responses to define levels of adequacy.

Thus the amino acid supply from microbial protein is calculated by multiplying the amino acid content of digestible microbial protein by the amount of microbial protein produced. The assumed amino acid proportions in digestible microbial amino acids are 0.0779, 0.0565, 0.0243, 0.0127 and 0.0175 for lysine, threonine, methionine, cystine and histidine respectively (derived from Rulquin *et al.,* 1998). That part of the MP derived from the digestible undegraded protein supply from each feed is multiplied by the amino acid content of that feed derived from the **FiM** feed database. The total supply of metabolisable amino acids (MAA) is calculated as the sum of amino acids from digestible microbial protein and from digestible undegraded dietary protein.

The principle adopted by INRA (Rulquin and Verité, 1993; Rulquin *et al.,* 1998) to calculate the supply of, and response to, amino acids is used. Broadly, this is based on the fact that a large body of research indicates that the dairy cow's requirement for total non-essential amino acids is met before the requirement for at least the most limiting of the essential amino acids. This being the case, then it follows that the efficiency of utilisation of MP for milk protein synthesis will be determined to a large degree by how well the profile of amino acids in the MP matches the profile required by the animal (so-called ideal protein). Thus the supply of individual MAA is expressed as a proportion of MP supply and the adequacy of the resultant values is interpreted in terms of milk protein output on the basis of a series of studies with dairy cows. Although the principle of the INRA approach was adopted, some modifications were however needed as the **FiM** model has variable outflow rates in contrast to the fixed values in the INRA model.

Calculation of supply and threshold levels for an expected response

Supply

The total MAA supply is calculated as the sum of the supply of digestible amino acids (DAA) from each feed plus that from microbial protein. To interpret the likely response in milk protein output, the MAA supply from each amino acid is expressed as g MAA/100 g total MAA (ie MP^{FIM}). Although this is done for each AA, only the values for lysine and methionine can be interpreted with confidence at present.

The supply of individual DAA from the rumen microbial protein (DAA_{micI}, g/d) is calculated as:

$$DAA_{micI} = DMTP^{FIM} \times micaa_i \qquad \text{(Equation 4.7)}$$

where $DMTP^{FIM}$ is the supply of digestible microbial true protein (g/d) and $micaa_i$ is the proportion of amino acid I in total microbial amino acids.

The supply of individual DAA from the rumen undegraded protein fraction of an individual feed (DAA_{feed}) is:

$$DAA_{dup\ I} = DUP^{FIM}_{feed} \times feedaa_i \qquad \text{(Equation 4.8)}$$

where DUP^{FIM}_{feed} is the supply of DUP^{FIM} from an individual feed (g/d), and $feedaa_i$ is the proportion of amino acid I in total amino acids in the feed.

If it is assumed that all of the digested AA is metabolised then the total supply of MAA is calculated as:

$$MAA_I = DAA_{micI} + DAA_{dup\ I} \qquad \text{(Equation 4.9)}$$

The concentration of total MAA_I in the MP^{FIM} supply (g/100g) from a feed is therefore:

$$[MAA_I] = 100 * (MAA_I) / MP^{FIM} \qquad \text{(Equation 4.10)}$$

The calculations are repeated for each feed in the total diet.

Corrections to relate predictions to observed supply

During development of the INRA approach, a comparison was made between the predicted values for intestinal AA composition with values measured in vivo. A database of duodenal AA flow was created from INRA research and the literature. This included 133 diets offered to dairy cows (84) and young cattle (49). In the statistical analysis, the effect of between-centre variability was removed. Regressions were then derived relating the model predictions to the observed values and these were applied to the predicted AA concentrations as a correction. The procedure is described by Rulquin *et al.* (1998) and Rulquin *et al.* (2001a) and the equations are:

Lysine	cmplys = mplys*0.759+1.904
Methionine	cmpmet = mpmet*0.733+0.322
Threonine	cmpthr = mpthr*0.546+2.387

(Equations 4.11)

No corrections are applied to cystine and histidine, c denotes corrected value

Over a range of dietary conditions, the application of these equations for lysine and methionine led on average to corrected values being 102 and 88% of the uncorrected values respectively.

Threshold values

The threshold values (near to asymptote) proposed by Rulquin *et al.* (2001a) to optimise milk protein output are adopted for the DSS (Figure 4.1). They were based on a large number of studies in which it was observed that milk protein output by cows in early lactation showed diminishing returns to increasing metabolisable lysine and methionine concentrations in the total MAA. The threshold values of 6.8 and 2.1 g/100 g MAA (MP[FIM]) for lysine and methionine respectively take into account both practical and economical considerations. Notably, these are close to the values proposed by NRC (2001).

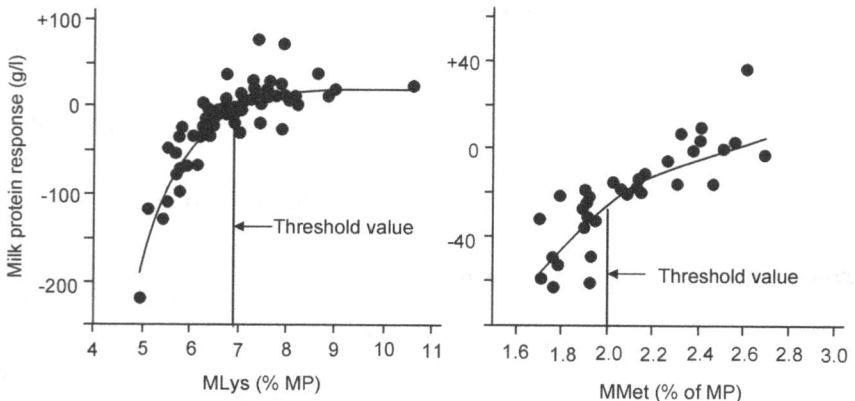

Figure 4.1. Response in milk protein to metabolisable lysine (MLys) and methionine (MMet) supply expressed as a percentage of total metabolisable protein (MP) supply (after Rulquin *et al.*, 2001b)

Assessment of adequacy of supply

The corrected concentrations of metabolisable lysine or methionine are then compared with the threshold values to assess the level of adequacy. If the concentrations do not reach the values, then the DSS will give a warning that the supply of lysine (or methionine) is marginal or deficient. Reformulation can then be undertaken to correct the situation.

Details of the equations used to calculate supply and adequacy are given on the accompanying CD. Amino acid concentrations of feeds are available from the feeds database and practical examples of ration formulation using the DSS are also shown on the CD.

A decision support system to predict the effect of diet composition on the composition of milk

Background

A major weakness of systems that predict the requirements and supply of energy and protein is their inability to predict effects of different diet compositions and intakes on milk composition. While it is recognised that the ultimate solution is a detailed mechanistic model to predict the response in milk components to changes in nutrient supply (TCORN Report No 11, 1998), the remit of **FiM** was to produce a requirement system. Nevertheless it is still feasible, within the context of a requirement system, to develop empirical models that can predict the direction of change in milk components. A decision support system (DSS) incorporating such models can be used to provide additional information in the formulation of rations.

Principles and methods

To be useful, a DSS incorporating empirical relationships should:

- be derived from data encompassing a wide range in diet and milk composition variables

- use sound statistical methods

- use variables that have biological meaning and

- produce an output that recognises the errors and limits of the methodology.

The approach was therefore to:

* construct a database

* produce multivariate models to predict milk composition

* convert models into an index to predict milk quality responses to diet change

* test the system as a predictor of response using data from the literature.

Database

A sub-set of eleven experiments (5 from ADAS, 4 from ARINI and 2 from SAC) was created from the intake database described in Chapter 1. The selection criteria were that experiments had a changeover design, contained an adequate description of diet formulation and that there were significant dietary effects on milk composition.

A wide variety of ration types was represented including diets based on grass silage, maize silage and including various by-product supplements and concentrate formulations. This resulted in a wide range of energy intakes and dietary concentrations of fibre, starch, sugar, protein and fat classes (e.g. saturated, unsaturated, etc.) that encompassed the main known dietary effects on milk composition. As expected, the variation in milk protein content was much less than for milk fat (CV% 5.0 and 13.1 respectively), although there was a similar degree of variation for protein and fat yields (CV% 18.2 and 17.0%).

Model construction

The final models were derived using the Partial Least Squares method and four milk composition / output indices were calculated:

* Fat Concentration (g/kg)

* Protein Concentration (g/kg)

* Fat Yield (g/d)

* Protein Yield (g/d)

Preliminary evaluation using multiple linear regression suggested that ten variables proved necessary to achieve acceptable predictions. Figure 4.2 below shows the weighted coefficients that take into account differences in the means and variances of the individual predictors. Thus, the graphs indicate the true relative influence of the predictors.

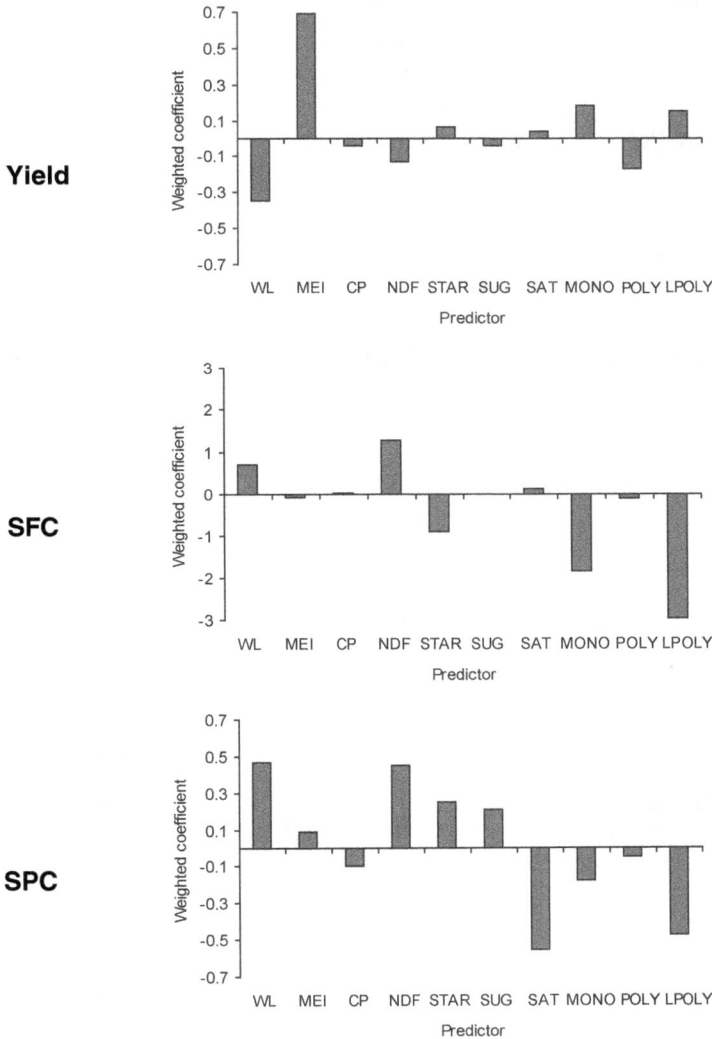

Yield

SFC

SPC

Figure 4.2 Weighted Coefficients describing responses in Milk Constituents (SFC, fat concentration; SPC, protein concentrations)

where WOL, cow week of lactation, MEI, ME intake (MJ/d) and Diet composition (g/kg DM) variables of [CP], crude protein;[NDF], neutral detergent fibre; [STARCH], starch; [SUGAR], sugar; [SAT], saturated fat; [MONO], mono-unsaturated fat; [POLY], long chain polyunsaturated fat; [LCPOLY], long-chain polyunsaturated fat (>= C20).

The factors that were determined to be significant in influencing milk composition and the direction of the coefficients accord with effects widely reported in the literature (Sutton, 1986). However, the positive coefficients for both NDF and STARCH for SPC appear surprising. These variables show a negative relationship to each other, ([NDF] = 445-0.58[STARCH] R^2 = 0.31), so that a change in [NDF] is typically associated with a much larger opposite change in [STARCH] to create the artifact.

Calculation of response indices

The aim of the DSS is to predict the *direction of response* to dietary change rather than absolute values. The initial models predicted yield and composition for any diet from the above variables. However, the addition of the mean values for yield, fat and protein content from the experiments allowed a better estimate of the coefficients that described the effect of *diet change* on yield and composition (i.e. *response*).

The equations to predict the indices are given below:

Predicted Milk yield (kg/d)

PREDY = 5.38 - .1427 * WL + .04849 * MEI - .001366 * [CPRAT] - .003657 * [NDFRAT] + .0001727 * [STARAT] - .0004392 * [SUGRAT] + .005951 * [SATRAT] + .06508 * [MONORAT] - .07675 * [POLYRAT] + .07395 * [LPOLYRAT] + .3501 * Y - .1361 * FAT + .2131 * PROT + .00095849 * MEI * Y

(Equation 4.11)

Predicted Milk fat concentration (g/kg)

[PREDFAT] = 32.59 - 0.02344 * WL + 0.010543 * MEI - 0.00441 * [CPRAT] + 0.008808 * [NDFRAT] - 0.00983 * [STARAT] - 0.01609 * [SUGRAT] + 0.081605 * [SATRAT] - 0.15172 * [MONORAT] - 0.01957 * [POLYRAT]- 1.78293 * [LPOLYRAT] - 0.18825 * Y + 0.490607 * [FAT] - 0.21412 * [PROT]

(Equation 4.12)

Predicted Milk protein concentration (g/kg)

[PREDPROT] = 8.04 - .01914 * WOL + .01688 * MEI - .00909 * [CPRAT] + .005348 * [NDFRAT]-.00249 * [MONORAT] + .016053 * [POLYRAT] - .40959 * [LPOLYRAT] - .10768 * Y - .00316 * FAT + .750456 * PROT

(Equation 4.13)

where WOL,cow week of lactation, MEI, ME intake (MJ/d) and total ration composition (g/kg DM) variables of [CPRAT], crude protein; [NDFRAT], neutral detergent fibre; [STARAT], starch; [SUGRAT], sugar; [SATRAT], saturated fat; [MONOTRAT], mono-unsaturated fat; [POLYRAT], long chain poly-unsaturated fat; [LPOLYRAT], long-chain poly-unsaturated fat (>= C20).

and Y, [FAT] and [PROT] are the base milk yield (kg/d), fat and protein concentrations derived from the current or historical average for the herd or group.

The indices are calculated from:

$$FATY = FAT * Y \qquad \text{(base fat yield)}$$
$$PROTY = PROT * Y \qquad \text{(base protein yield)}$$
$$PREDFATY = PREDY * PREDFAT \qquad \text{(predicted fat yield)}$$
$$PREDPROTY = PREDY * PREDPROT \qquad \text{(predicted protein yield)}$$

$$FATindex = PREDFAT - FAT \qquad \text{(fat concentration index)}$$
$$FATYindex = PREDFATY - FATY \qquad \text{(fat yield index)}$$
$$PROTindex = PREDPROT - PROT \qquad \text{(protein concentration index)}$$
$$PROTYindex = PREDPROTY - PROTY \qquad \text{(protein yield index)}$$

Evaluation of the DSS

A test of the system was carried out using data from ten changeover experiments reported in the literature. For each experiment, indices were calculated for the control diet and for each of the treatment diets. This allowed the predicted responses to be compared with the measured responses. The experiments ranged over studies of the effects of varying level and quality of forage and supplement including the use of oil supplements. The results are shown in Figure 4.3.

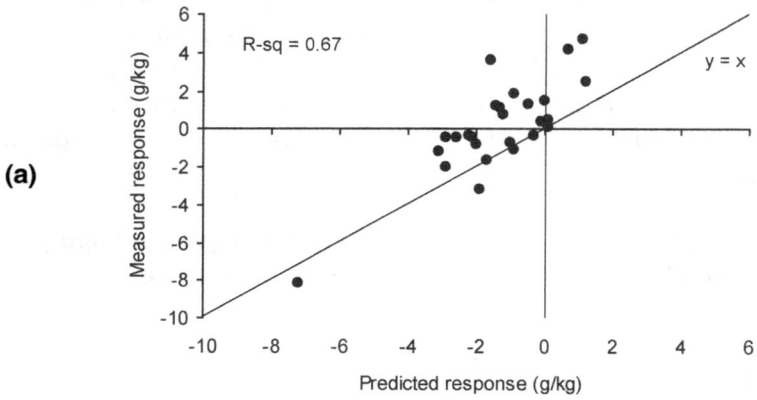

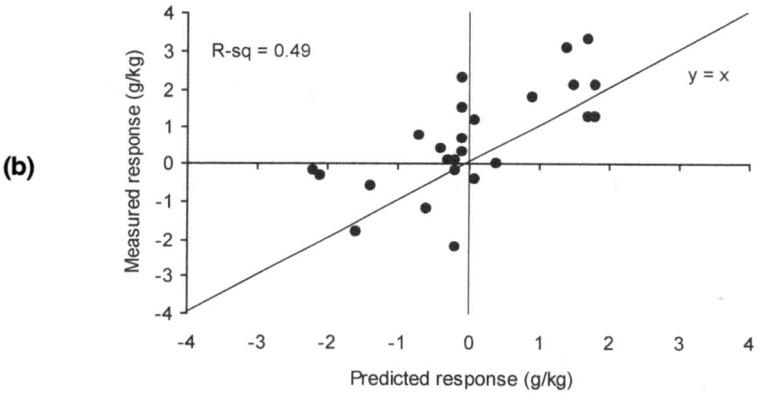

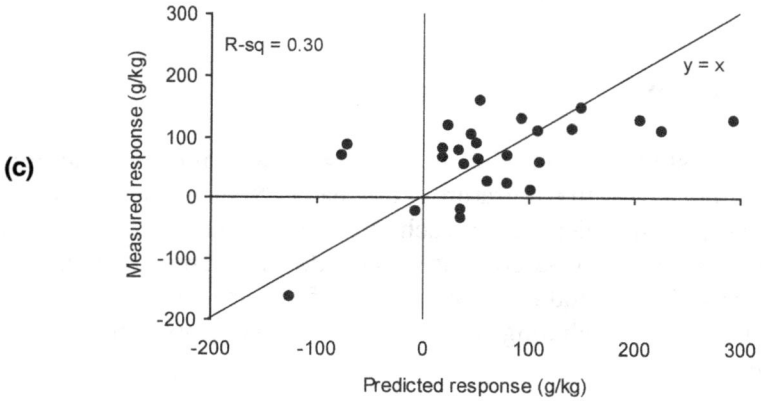

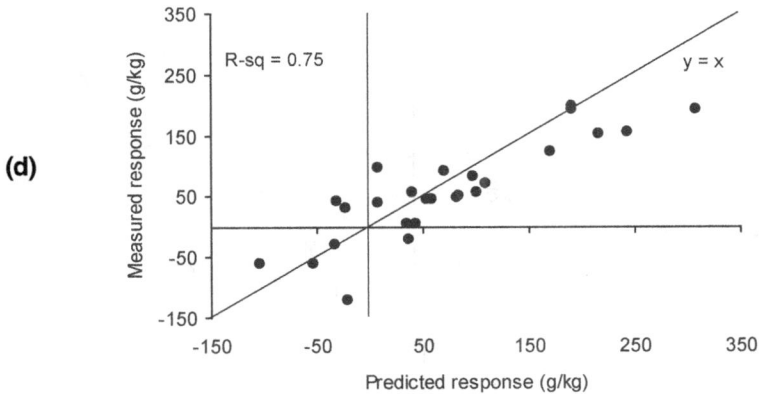

(d)

Figure 4.3 Comparison of measured and predicted responses
a) Fat concentration; b) Protein concentration; c) Milk fat yield; d) Milk protein yield

The DSS under-predicted the response for milk fat content by 1.5 g/kg and milk protein content by 0.6 g/kg. There were also small biases in the prediction of fat and protein yield responses. The R^2 value for the protein yield relationship (0.75) was much higher than for fat yield (0.30) suggesting that the errors in yield and protein content response prediction are not additive i.e. the responses of volume and protein content are linked. An example of this is the observation that, in some experiments, cows respond to dietary change by increasing yield without a change in protein content (an increase in protein yield) whilst, in others, the same yield response is cause by an increase in protein content at the same yield.

Accurate prediction of responses to dietary change is not easy to achieve. The accuracy of the empirical methods shown here appears adequate to provide general guidance on the likely response of milk output to dietary manipulation. The prediction of protein yield response appears remarkably accurate in the validation exercise. The ability of the DSS to predict the consequences of changes in the level and type of oil in the diet is particularly useful. The DSS should make a useful addition to rationing software by giving guidance as to the likely outcome of new diets.

Interpretation of response

This set of calculations to calculate the fat and protein indices (see above) for the initial diet is then repeated for a new diet (diet 2) *using the same target values for Y, [FAT] AND [PROT]*. Figure 4.4 shows the framework to calculate response.

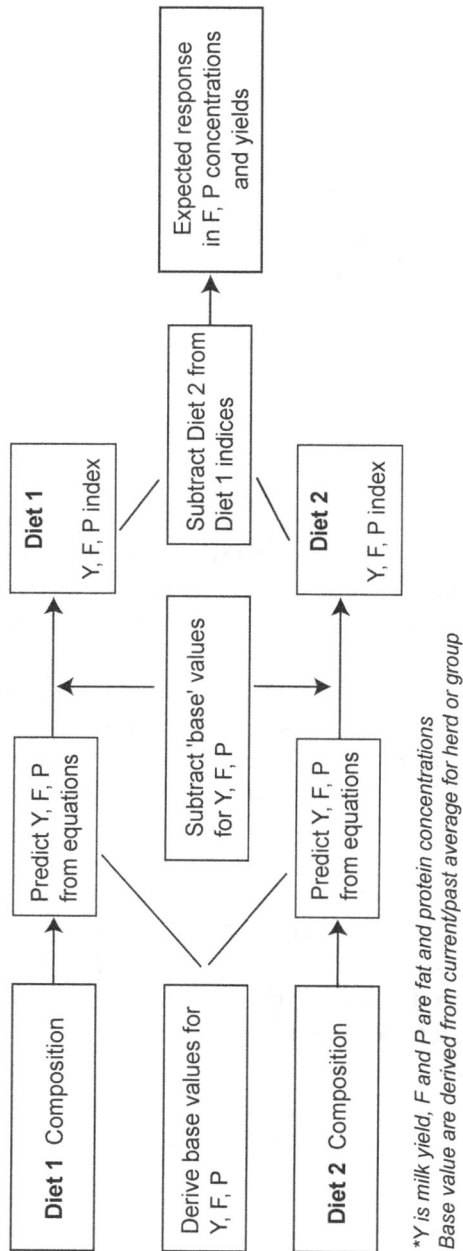

Figure 4.4 Framework for Milk Composition DSS

Predicted responses are then calculated:

predicted fat concentration response
$$\text{FATRESPONSE} = \text{FATindex}_{\text{DIET 1}} - \text{FATindex}_{\text{DIET 2}}$$

predicted fat yield response
$$\text{FATYRESPONSE} = \text{FATYindex}_{\text{DIET 1}} - \text{FATYindex}_{\text{DIET 2}}$$

predicted protein concentration response
$$\text{PROTRESPONSE} = \text{PROTindex}_{\text{DIET 1}} - \text{PROTindex}_{\text{DIET 2}}$$

predicted protein yield response
$$\text{PROTYRESPONSE} = \text{PROTYindex}_{\text{DIET 1}} - \text{PROTYindex}_{\text{DIET 2}}$$

To enable the direction and extent of change to be described, the responses in fat and protein (concentration and yield) were derived from the database that was used to generate the prediction equation and divided into quartiles. The output from the DSS then shows the direction of change (or no change) and the extent of change based on the quartile ranges (small, moderate large, very large).

The use of these indices in diet formulation and practical examples of rations derived using the indices are given on the accompanying CD.

Conclusions

The accuracy and precision of the empirical methods used in the DSS are adequate to provide general guidance on the likely response of milk component output to dietary manipulation. As such, the DSS makes a useful addition to rationing software by giving guidance as to the likely outcome of new diets. The prediction of protein yield response appears remarkably accurate in the validation exercise. Further, the ability of the DSS to predict the consequences of changes in the level and type of oil in the diet is particularly useful.

5   Summary of feed characterisation methods

C. Rymer and R.E. Agnew

A comprehensive range of feed characterisation methodology supports the **Feed into Milk** system and an extensive database of feed values is provided on the CD accompanying this book. This database together with the complete list of equations in software format, also on the CD, enables the user to begin formulating diets with almost immediate effect.

New reference methods for the feed characteristics were derived where necessary and these together with existing methods were used to develop NIRS (near infra red spectroscopy) predictive methods for forages in their fresh state. The methods are summarised in Table 5.1.

The NIRS predictions, in the main, apply to grass silages although new calibrations are being developed for other forages. They are intended for large-scale use in commercial analytical laboratories. The matching of NIRS instruments across these laboratories and quality control through ring tests is a vital part of the scheme and this is undertaken through the Forage Analytical Assurance Group (FAA). The FAA group also commission research to advance the analytical and predictive methods and to expand the range of forages in the database.

Access to the methodologies and quality assurance system is through the Technical Secretary of FAA, Dr J.S. Blake (jsb@jsblake.co.uk).

Table 5.1 Summary of feed characterisation methods
a) Estimation of intake and metabolisable energy supply

Characteristic	Reference method	Prediction method
Dry matter (DM, g/kg fresh weight)	AOAC (1995)	
Toluene dry matter (TDM, g/kg fresh weight)	MAFF (1986)	
Forage intake potential (FIP, forages, g/kg $W^{0.75}$)	Agnew *et al.* (2001)	Grass silage: NIRS Non grass silage with DM<550 g/kg: Non fermented: FIP=(0.122DM+0.088DOMD) + 3.84) or Fermented: FIP = 0.122 DM + 0.088 DOMD + 14.92 pH – 0.076 NH$_3$ – 51.3 Non grass silage with DM>550 g/kg: FIP= (8 / (1 – (.01 * ME / .15))) * (W / $W^{.75}$) where W is cow liveweight (kg) – adapted from Lewis (1981)
Starch (forages, g/kg DM)	AOAC (1995)	
Crude protein (CP, g/kg DM)	AOAC (1995)	
Metabolisable energy (ME, MJ/kg DM)	Rymer (2000) Park *et al.* (1998)	Grass and maize silage: NIRS prediction of digestible organic matter in the dry matter (DOMD) and that ME=0.016 DOMD. Compound feeds: ME=0.014 NCGD[1]+0.025 AHEE
[1]Neutral detergent cellulase plus gamannase digestibility (NCGD; compound feeds, g/kg DM)	MAFF (1993)	
Acid hydrolysed ether extract (AHEE; compound feeds, g/kg DM)	MAFF (1993)	

Table 5.1 Contd.

Characteristic	Reference method	Prediction method
b) Estimation of metabolisable protein supply		
Acid detergent insoluble N (ADIN, *g/kg DM*)	Goering *et al.* (1972)	
Soluble N	Weisbjerg *et al.* (1990)	Grass silage: NIRS
Soluble dry matter	Weisbjerg *et al.* (1990)	Grass silage: NIRS
Degradability characteristics *a,b* and *c* for dry matter	AFRC (1992); Ørskov and McDonald (1979)	Grass silage: NIRS prediction of DM losses after various incubation periods plus curve fitting software.
Degradability characteristics *a,b* and *c* for nitrogen	AFRC (1992); Ørskov and McDonald (1979)	Grass silage: NIRS prediction of N losses after various incubation periods plus curve fitting software.
ATP yield (mol/kg DM apparently degraded)	24 h incubation of dried, ground feed with buffered rumen fluid followed by filtering to determine apparent degradability of dry matter, and measurement of short chain (SCFA) fatty acid concentration of filtrate. ATP production calculated from net production of SCFA.	ATP yield = 27.3 - 0.0249 CP (g/kgDM) R^2(adj)= 0.567, rsd=3.73 *where CP is crude protein content of feed.*
c) Estimation of rumen stability value		
Characteristic	*Reference method*	*Prediction method*
Potential acid load (PAL, meq/kg DM)	24 h incubation of feed with buffered rumen fluid. Concentrate feeds are dried and ground but forages are milled after freezing in liquid nitrogen. PAL is a measure of the amount of alkali needed to raise pH of the incubation mixture back to pH 7.25.	Grass silage: NIRS Maize silage: PAL= 1691-2.31 TDM

References

AFRC, 1990. Technical Committee on Responses to Nutrients, Report Number 5, Nutritive Requirements of Ruminant Animals: Energy. *Nutrition Abstracts and Reviews, Series B,* **60**: 729-804.

AFRC, 1992. Technical Committee on Responses to Nutrients, Report N. 9. Nutritive Requirements of Ruminant Animals: Protein. *Nutrition Abstracts. and Reviews, Series B,* **62**:787-835.

AFRC, 1993. Energy and Protein Requirements of Ruminants. An advisory manual prepared by the AFRC Technical Committee on Responses to Nutrients. CAB International, Wallingford, UK.

AFRC, 1998. Response in the Yield of Milk Constituents to the Intake of Nutrients by Dairy Cows. AFRC Technical Committee on Responses to Nutrients, Report No 11. CABI Publishing, Wallingford, UK.

Agnew, R.E. and Yan, T., 2000. Impact of recent research on energy feeding systems for dairy cattle. *Livestock Production Science,* **66**: 197-215.

Agnew, R.E., Offer, N.W., McNamee, B.F. and Park, R.S., 2001. The development of a system based on near infrared spectroscopy to predict the intake of grass silage as the sole feed by the dairy cow. In: The Proceedings of Annual Meeting of British Society of Animal Science, 2001. p 197.

AOAC, 1995. Official methods of analysis of AOAC International. 16th edition, AOAC International 1995, Arlington, VA, USA.

ARC, 1965. The Nutrient Requirements of Farm Livestock. No. 2, Ruminants. HMSO, London, UK .

ARC, 1980. Nutrient Requirements of Ruminant Livestock. Technical review by an Agricultural Research Council Working Party. Commonwealth Agricultural Bureau, Farnham Royal, Slough, UK.

Archimède, H., Sauvant, D. and Schmidely, P., 1997. Quantitative review of ruminal and total tract digestion of mixed diet organic matter and carbohydrates. *Reproduction Nutrition Development,* **37**: 173-189.

Beever, D.E., 1993. Characterisation of forages: appraisal of current practice and future opportunities. In: *Recent Advances in Animal Nutrition 1993* (Eds. P.C. Garnsworthy and D.J.A. Cole). Nottingham University Press,

Nottingham, UK. pp 1-17.

Blaxter, K.L., 1962. The energy metabolism of ruminants. Hutchinson, London, UK.

Burroughs, W., Nelson, D.K. and Mertens, D.R., 1975. Protein physiology and its application in the lactating dairy cow: The metabolisable protein feeding standard. *Journal of Agricultural Science*, **41**: 933-944.

Cammell, S.B., Dhanoa, M.S., Beever, D.E., Sutton, J.D. and France, J., 1998. An examination of the metabolisble energy requirement of lactating dairy cows. In: The Proceedings of Annual Meeting of British Society of Animal Science, p. 197, Scarborough, England.

CSIRO, 1990. Feeding Standards for Australian Livestock. Ruminants. CSIRO, Victoria, Australia.

De Brabander, D.L., De Boever, J.L., De Smet, A.M., Vanacker, J.M. and Boucqué, C.V., 1996. Evaluation of physical structure in dairy cattle nutrition. Condensed version. Report of the Study Centre of Dairy Cattle Husbandry, Communication No. 990.

Dhanoa, M.S., France, J., López, S., Dijkstra, J., Lister, S.J., Davies, D.R. and Bannink, A., 1999. Correcting the calculation of extent of degradation to account for particulate matter loss at zero time when applying the polyester bag method. *Journal of Animal Science*, **77**: 3385-3391.

Dulphy, J.P., Demarquilly, C. and Remond, B., 1989. Quantities d'herbe ingerees par les vaches laitieres, les genisses et les moutons: effet de quelques facteurs de variation et comparison autre ces types d'animaux. *Annales de Zootechnie*, **38**: 107-120.

Givens, D.I. and Gill, M., 1998. Current and future potential of alternative techniques. In: E.R. Deaville, E. Owen, A.T. Adesogan, C. Rymer, J.A. Huntington and T.L.J. Lawrence (Editors), *In vitro techniques for measuring nutrient supply to ruminants*. British Society of Animal Science Occasional Publication No. 22. pp 161-171.

Goering, H.K., Gordon, C.H., Hemken, R.W., Waldo, D.R., van Soest, P.J. and Smith, L.W., 1972. Analytical estimates of nitrogen digestibility in heat damaged forages. *Journal of Dairy Science*, **55**: 1275-1280.

Hespell, R.B., 1979. Efficiency of growth by ruminal bacteria. *Federation Proceedings*, **38**: 2707-2712.

Hvelplund, T. and Weisbjerg, M., 2000. *In situ* techniques for the estimation of protein degradability. In: Givens, D.I., Owen, E., Axford, R.F.E. and Omed, H.M. (Eds) *Forage Evaluation in Ruminant Nutrition*, CABI Publishing, Wallingford, UK.

Kebreab, E., France, J., Agnew, R.E., Yan, T., Dhanoa, M.S., Dijkstra, J., Beever, D.E. and Reynolds, C.K., 2003. Alternative to linear analysis of energy balance data from lactating dairy cows. *Journal of Dairy Science*, **86**: 2904-2913.

Lewis, M.,1981. Equation for predicting silage intake by beef and dairy cattle. In: Proceedings of the Sixth Silage Conference, Edinburgh, pp.35-36.

Madsen, J., 1985. The basis for the Nordic protein evaluation system for ruminants. The AAT/PBV system. *Acta Agriculturae Scandinavica.* Suppl **25**: 9-20.

MAFF, 1975. Energy allowances and feeding systems for Ruminants. MAFF Technical Bulletin No. 33. HMSO, London, UK.

MAFF, 1986. The analysis of agricultural materials. Third edition. Reference Book 427, HMSO, London, UK. pp.85-87.

MAFF, 1993. Prediction of energy value of compound feedingstuffs for farm animals. Booklet 1285, MAFF Publications, Alnwick, UK.

Mayne, C.S., 1990. The effect of protein content of the supplement on the response to level of supplementation with dairy cattle offered grass silage *ad libitum*. *Animal Production*, **50**: 556 (Abs).

McBurney, M.I., Van Soest, P.J. and Chase, L.E., 1983. Cation exchange capacity and buffering capacity of Neutral Detergent Fibres. *Journal of the Science of Food and Agriculture*, **34**: 910-916.

Metcalf, J.A., Mansbridge, R.J, Blake, J.S., Newbold, J.R. and Oldham, J.D., 1997. Examining the metabolisable protein system at ideal protein:energy ratios in dairy cows. In: The Proceedings of Annual Meeting of British Society of Animal Science 1997, p83.

Milligan, R.A., Chase, L.E., Sniffen, C.J. and Knoblanch, W.A., 1981. Least-cost ration balanced dairy rations. A computer program users manual. Animal Science, Mimeos, 54 Cornell University, Ithaca, NY, USA.

Newbold, J.R., Cottrill, B.R., Mansbridge, R.J. and Blake, J.S., 1994. Effects of metabolisable protein on intake of grass silage and milk protein yield in dairy cows. *Animal Production*, **58**: 455 (Abs).

NRC, 1985. Ruminant Nitrogen Usage. National Academy Press, Washington DC, USA.

NRC, 1988. Nutrient requirements of dairy cattle. Sixth revised edition, National Academy Press, Washington, DC, USA.

NRC, 2001. Nutrient Requirements of Dairy Cattle, Seventh Revised Edition. National Academy Press, Washington, DC, USA.

Oldham, J.D., Emmans, G. and Friggens, N., 1998. Development of predictive systems to relate animal and feed characteristics to amounts and patterns of forage and total food intake by cows. In: *Optimisation of forage quality and intake by ruminants*. MAFF RUMINT Report DS04.

Ørskov, E.R. and Mehrez, A.Z., 1977. Estimation of extent of protein degradation from basal feeds in the rumen of sheep. *Proceedings of the Nutrition Society,* **36**: 78A.

Ørskov, E.R. and McDonald, I., 1979. The estimation of protein degradability in the rumen from incubation measurements weighted according to rate of passage. *Journal of Agricultural Science,* **92**: 499-503.

Ørskov, E.R., MacLeod, N.A. and Kyle, D.J., 1986. Flow of nitrogen from the rumen and abomasum in cattle and sheep given protein-free nutrients by intragastric infusion. *British Journal of Nutrition,* **56**: 241-248.

Osbourn, D.F. and Thomas, C., 1989. The feeding value of grass and grass products, In: W Holmes (ed) *Grass – Its Production and Utilization.* British Grassland Society. UK.

Owens, F.N. and Goetsch, A.L., 1986. Digesta passage and microbial protein synthesis. In: Milligan, L.P., Grovum, W.L. and Dobson, A. (Eds). Control of digestion and metabolism in ruminants. Prentice-Hall, NJ, USA. pp 196-223.

Park, R.S., Agnew, R.E., Gordon, F.J., Steen., R.W.J., 1998. The use of near infrared reflectance spectroscopy (NIRS) on undried samples of grass silage to predict chemical composition and digestibility parameters. *Animal Feed Science and Technology,* **72**: 155-167.

Rulquin H., Guinard J. and Vérité R., 1998. Variation of amino acid content in the small intestine digesta of cattle: Development of a prediction model. *Livestock Production Science,* **53**: 1-13.

Rulquin, H., and Vérité R., 1993. Amino acid nutrition of dairy cows: Productive effects and animal requirements. In: P. C. Garnsworthy and D. J. A. Cole, (Eds) *Recent Advances in Animal Nutrition 1993.* Nottingham University Press, Nottingham, UK. pp. 55-77.

Rulquin, H., Vérité R and Guinard-Flament, J., 2001a. Acides amines digestible dans l'intestine. Le système AADI et les recommendations d'apport pour la vache laitière. *INRA Productions Animales,* **14**: 265-274.

Rulquin, H., Vérité R., Guinard-Flament, J. and Pisulewski, P.M., 2001b. Acides amines digestible dans l'intestine. Origines des variations chez les ruminants et répercussions sur les protéines du lait . *INRA Productions Animales,* **14**: 201-210.

Russell, J.B., O'Connor, J.D., Fox, D.G., Van Soest, P.J. and Sniffen, C.J,. 1992. A net carbohydrate and protein system for evaluating cattle diets. I. Ruminal fermentation. *Journal of Animal Science,* **70**: 3551-3561.

Rymer, C., 2000. The measurement of forage digestibility in vivo. In: D I Givens, E Owen, R F E Axford and Omed H M (Eds), *Forage Evaluation in Ruminant Nutrition,* CABI Publishing, Wallingford, UK. pp. 113-134.

Sanderson, R., Lister, S.J., Sargeant, A. and Dhanoa, M.S., 1997. Effect of particle size on in vitro fermentation of silages differing in dry matter content. In: The Proceedings of Annual Meeting of British Society of Animal Science 1997, p. 197.

Sauvant, D. and Archimède, H., 1989. The prediction of the digestion passage rate in ruminants. Document interne à la station de nutrition et alimentation, INRA-IAPG, France.

Sniffen, C.J., O'Connor, J.D, Van Soest, P.J., Fox, D.G. and Russell, J.B., 1992. A net carbohydrate and protein system for evaluating cattle diets. II. Carbohydrate and protein availability. *Journal of Animal Science*, **70**: 3562-3577.

Steen, R.W.J., Gordon, F.J., Dawson, L.E.R., Park, R.S., Mayne, C.S., Agnew, R.E., Kilpatrick, D.J. and Porter, M.G., 1998. Factors affecting the intake of grass silage by cattle and prediction of silage intake. *Animal Science*, **66**: 115-127.

Tamminga, S., van Straalen, W.M., Subnel, A.P.J., Meijer, R.G.M., Steg, A., Wever, C.J.G. and Blok, M.C., 1994. The Dutch protein evaluation system: the DVE/OEB-system. *Livestock Production Science*, **40**: 130-155.

Tuori, M., Kaustell, K.V. and Huhtanen, P., 1998. Comparison of the protein evaluation systems for feeds for dairy cows. *Livestock Production Science*, **55**: 33-46.

Tyrrell, H.F. and Reid, J.T., 1965. Prediction of the energy value of cow's milk. *Journal of Dairy Science*, **48**: 1215-1223.

Vadiveloo, J. and Holmes, W., 1979. The prediction of the voluntary feed intake of dairy cows. *Journal of Agricultural Science*, **93**: 553-562

Van Es, A.J.H., 1978. Feed evaluation for ruminants. 1. The systems in use from May 1977 onwards in the Netherlands. *Livestock Production Science*, **5**: 331-345.

van Straalen, W.M., Salaün, C., Veen, W.A.G., Rijekpma, Y.S., Hof, G. and Boxem, T.J., 1994. Validation of protein evaluation systems by means of milk production experiments with dairy cows. *Netherlands Journal of Agricultural Science*, **42**: 89-104.

Vérité, R., Michalet-Doreau, B., Chapoutot, P., Peyraud, J-L., and Poncet, C., 1987. Révision du système des protéines distibles dans l'intestin (PDI). Bulletin Technique C.R.Z.V. Thiex. INRA. **70**: 19-34.

Weisbjerg, M.R., Bhargava, P.K. and Hvelplund, T.A., 1990. Use of degradation curves in feed evaluation (translated version). In: Report from the National Institute of Animal Science, Denmark, Beretning Fra Statens Husdyrbqugsforog, Denmark.

Yan, T., Gordon, F.J., Agnew, R.E., Porter, M.G. and Patterson, D.C., 1997a.

The metabolisable energy requirement for maintenance and the efficiency of utilisation of metabolisable energy for lactation by dairy cows offered grass silage-based diets. *Livestock Production Science*, **51**: 141-150.

Yan, T., Gordon, F.J., Ferris, C.P., Agnew, R.E., Porter, M.G. and Patterson, D.C., 1997b. The fasting heat production and effect of lactation on energy utilization by dairy cows offered forage-based diets. *Livestock Production Science*, **52**: 177-186.

Yan, T., Agnew, R.E. and Gordon, F.J., 2002. The combined effects of animal species (sheep versus cattle) and levels of feeding on digestible and metabolisable energy concentrations in grass silage-based diets of cattle. *Animal Science*, **75**: 141-151.

Glossary – definition of terms used in equations

This glossary is for use with Chapters 1 to 5. It follows the convention adopted in AFRC (1993). Terms in:

- upper case represent supply in g or MJ/d, e.g. EDN, supply of effective degradable N (g/d)

- [ ] are a concentration e.g. [CP], concentration of crude protein in a feed (g/kgDM)

- lower case indicate a proportion, rate or efficiency e.g. a, DM fraction lost instantaneously *in situ* (proportion)

A subscript denotes a function for a supply or requirement or a qualification of a supply or concentration e.g. ATP_{ssp}, supply of ATP from the soluble and small particle fractions of the feed (mols/d).

The superscript FiM is used to identify terms terms that have been used in previous UK models that are numerically different in the **FiM** system.

[ADIN]	Concentration of acid detergent insoluble N (g/kgDM)
ATP_{lp}	Supply of ATP from the large particle fractions of the feed (mols/d)
ATP_{ssp}	Supply of ATP from the soluble and small particle fractions of the feed (mols/d)
ATPy	Yield of ATP (mol per kg of DM degraded)
a	DM fraction lost instantaneously *in situ* (proportion)
a_N	N lost instantaneously *in situ* (proportion)
b	DM potentially degraded *in situ* (proportion)
$ß_N$	N potentially degrade *in situ* (proportion)
$ß_D$	Degradable small particle DM fraction of a feed (proportion)
$ß_{DN}$	Degradable small particle N fraction of a feed (proportion)

c	Rate constant for degradation of 'b' DMfraction *in situ* (proportion/h)
c_N	Rate constant for degradation of 'b' N fraction (proportion/h)
CDMI	Concentrate dry matter intake (kg/day)
[CCP]	Concentration of crude protein in the concentrate (g/kg total concentrate DM
[CP]	Concentration of crude protein in a feed (g/kgDM)
[CPRAT]	Concentration of crude protein in a ration (g/kgDM)
CS	Condition score of the cow (1 – 5 scale)
DAA_{dupI}	Supply of digestible undegraded amino acid I from a feed (g/d)
DAA_{micI}	Supply of digestible microbial amino acid I from a feed (g/d)
[DM]	Dry matter concentration (g/kg fresh weight)
DMI, TDMI	Dry matter intake (kg/day)
$DMTP^{FIM}$	Supply of digestible microbial true protein (g/d)
DUP^{FIM}	Supply of digestible undegraded protein (g/d)
DUP^{FIM}_{feed}	Supply of DUP^{FIM} from an individual feed (g/d)
$eddm_{lp}$	Effective degradability of the large particle fraction of the DM (proportion)
$eddm_{ssp}$	Effective degradability of the soluble and small particle fractions of the DM (proportion)
edn	Effective degradability of the N fraction of a feed (proportion)
EDN	Supply of effective degradable N
E_l	Milk energy output (MJ/cow/day)
$E_{l(0)}$	Milk energy output adjusted for body weight change (MJ/kgW$^{.75}$)

$E_l corr$	Milk energy yield corrected for weight loss (MJ/kgW$^{0.75}$)
E_{lWC}	Net energy used for milk production from weight loss (MJ/d)
EV_g	Net energy value of weight change (MJ/kg)
EV_l	Energy value of milk (MJ/kg)
f	Forage DM in the diet (proportion)
[FAT]	Milk fat concentration (g/kg)
FE	Energy lost in faeces (MJ/d)
feedaa$_I$	Proportrion of amino acid I in total amino acids in a feed (proportion)
FIP	Forage intake potential (g DM/kg W$^{.75}$)
[FS]	Forage starch concentration (g/kg DM)
GE	Gross energy (heat of combustion) of the feed consumed (MJ/kgDM)
k_c	Fractional outflow rates of concentrates (proportion/h)
k_f	Fractional outflow rates of forages (proportion/h)
k_{liq}	Fractional outflow rates of liquids (proportion/h)
k_g^{FiM}	Efficiency of utilisation of ME for gain (proportion)
k_l^{FiM}	Efficiency of utilization of ME for lactation (proportion)
k_{nl}	Efficiency of utilization of metabolisable protein for milk protein synthesis (proportion)
k_t	The efficiency with which body energy is used to support milk production for cows in negative energy balance (proportion)
[LPOLYRAT]	Concentration of long-chain poly-unsaturated fat (>= C20) in a ration (g/kgDM)
M_{act}	ME for activity (MJ/d)
M_c	ME required for pregnancy (MJ/d)

Feed into Milk

M_g^{FiM}	ME required for weight gain (MJ/d)
M_m	ME required for maintenance (MJ/kg $W^{0.75}$)
M_{ml}	ME required for maintenance and milk production (MJ/kg$W^{0.75}$)
MAA_I	Supply of metabolisable amino acid I from a ration (g/d)
$[MAA_I]$	Concentration of metabolisable amino acid I in the metabolisable protein of the total ration (g/100gMP^{FiM})
MCP_{atp}	Microbial crude protein supply limited by ATP supply(g/d)
MCP_{edn}	Microbial crude protein supply limited by EDN supply (g/d)
MCP^{FIM}	The actual microbial crude protein supply (g/d) i.e. the lower of MCP_{atp} or MCP_{edn}
MDM	Yield of microbial DM (g /d))
Mealfac	Meal frequency factor
MEI	Metabolisable energy intake(MJ or MJ/kgW$^{.75}$)
MethE	Energy lost in methane (MJ/d)
$[ME_m]$	Concentration of ME (MJ/kgDM) measured at the maintenance level of feeding
$[ME_p]$	Concentration of ME (MJ/kgDM) measured at the production level of feeding
M_{req}^{FIM}	Total ME requirement (MJ/d)
MFP	Metabolic faecal protein (g/d)
$micaa_I$	Proportion of amino acid I in total microbial amino acids (proportion)
[MONOTRAT]	Concentration of mono-unsaturated fat in a ration
MP^{FIM}	Metabolisable protein supply (g/d)
MP_{req}^{FIM}	Metabolisable protein requirement (g/d)
MP_c	Metabolisable protein required for pregnancy (g/d)
MP_g	Metabolisable protein required for weight gain (g/d)

MP_l	Metabolisable protein required for lactation(g/d)
MP_{loss}	Metabolisable protein derived from weight loss (g/d)
MP_m^{FIM}	Metabolisable protein required for maintenance (MJ/d)
MP_c	Metabolisable protein required for pregnancy post 250 days (MJ/d)
[N]	Concentration of N in a feed (g/kgDM)
[NDF]	Concentration of neutral detergent fibre in a feed (g/kgDM)
[NDFRAT]	Concentration of neutral detergent fibre in a ration (g/kgDM)
NE	Net energy (MJ/d)
PAL	Potential acid load (meq/kgDM)
[POLYRAT]	Long chain poly-unsaturated fat in ration (g/kgDM)
[PREDFAT]	Predicted milk fat concentration (g/kg)
[PREDPROT]	Predicted milk protein concentration (g/kg)
PREDY	Predicted yield (kg/d)
[PROT]	Milk protein concentration (base, g/kg)
[RSV]	Rumen stability value for a ration
$[RSV_{req}]$	Requirement for rumen stability value
RSV_{feed}	Supply of RSV from a feed
RSV_{supply}	Supply of RSV from a ration
$RSV_{balance}$	RSV balance for ration
s	Soluble DM fraction (proportion)
s_{corr}	Soluble DM corrected fat fermentation acids (proportion)
s_N	Soluble N fraction (proportion)
[SATRAT]	Concentration of saturated fat in ration (g/kgDM)
$SDMI_0$	Silage DM intake at zero concentrate intake (g DM/kg$^{0.75}$)
$SDMI_c$	Silage DM intake at concentrate intake c (g DM/kg$^{0.75}$)

Feed into Milk

[STARAT]	Concentration of starch in a ration (g/kgDM)
[SUGRAT]	Concentration of sugar in a ration (g/kgDM)
[TFA]	Total fermentation acids (g/kgDM)
UDP	Supply of undegraded protein (g/d)
WC	Weight change (kg/d)
WOL	Week of lactation
Y	Milk yield (kg/d)
Y_{ATPlp}	Microbial efficiency for large particles (g microbial dry matter/mol ATP)
Y_{ATPssp}	Microbial efficiency for soluble and small particles (g microbial dry matter/mol ATP)

www.ingramcontent.com/pod-product-compliance
Lightning Source LLC
Chambersburg PA
CBHW061300220326
41599CB00028B/5724